AGRARIAN CAPITALISM, WAR AND PEACE IN COLOMBIA

Based on extensive research conducted in Colombia since 2009, this book addresses the connection between land grabbing and agrarian capitalism, as well as the unfulfilled promises of peace and justice.

While land remains a key resource at the core of many contemporary civil wars, the impact of high-intensity armed violence on the formation of agrarian capitalism is seldom discussed. Drawing on nearly 200 interviews, archival research, and geographical data, this book examines land grabbing and the role of violence in capital with a particular focus on one key actor in the Colombian civil war: paramilitary militias. This book demonstrates how the intricate ties between armed conflict and economy formation are obscured by the widespread belief that violence is a radical form of action, breaking with the normal course of society and disconnected from the legal economy. Under this view, dispossession is perceived as diametrically opposed to capitalist accumulation. This belief is enormously influential in precisely those bureaucratic agencies that are in charge of peacebuilding, both domestically and internationally. However, this narrow view of the relationship between armed violence and capitalism belies the close ties between plunder and lawful profit, and obscures the continuity between violent dispossession and the free market. By the same token, it legitimizes post-war inequality in the name of capitalist development. The book concludes by arguing that the promotion of radical democracy in the government of land and rural development emerges as the only reasonable path for pacifying a violent polity.

The book is essential reading for students, scholars, and development aid practitioners interested in land and resource grabbing, agrarian capitalism, civil wars, and conflict resolution.

Jacobo Grajales is Professor of Political Science at the University of Lille, France, and a Fellow of the Institut Universitaire de France (IUF).

ROUTLEDGE STUDIES IN GLOBAL LAND AND RESOURCE GRABBING

This series presents and discusses 'resource grabbing' research in a holistic manner by addressing how the rush for land and other natural resources (water, forests, minerals, etc.) is intertwined with agriculture, mining, tourism, energy, carbon markets, climate change, and disasters. This series welcomes contributions from a wide range of interdisciplinary approaches and on a global basis.

Series Editors:
Andreas Neef, The University of Auckland, New Zealand
Chanrith Ngin, The University of Auckland, New Zealand

Titles in this series include:

Industrial Tree Plantations and the Land Rush in China
Implications for Global Land Grabbing
Yunan Xu

Capitalism and the Commons
Just Commons in the Era of Multiple Crises
Edited by Andreas Exner, Sarah Kumnig and Stephan Hochleithner

Tourism, Land Grabs and Displacement
The Darker Side of the Feel-Good Industry
Andreas Neef

Agrarian Capitalism, War and Peace in Colombia
Beyond Dispossession
Jacobo Grajales

For more information about this series, please visit: www.routledge.com/sustainability/series/GLRG

AGRARIAN CAPITALISM, WAR AND PEACE IN COLOMBIA

Beyond Dispossession

Jacobo Grajales

Routledge
Taylor & Francis Group
LONDON AND NEW YORK

earthscan
from Routledge

First published 2021
by Routledge
2 Park Square, Milton Park, Abingdon, Oxon OX14 4RN

and by Routledge
605 Third Avenue, New York, NY 10158

Routledge is an imprint of the Taylor & Francis Group, an informa business

British Library Cataloguing-in-Publication Data
A catalogue record for this book is available from the British Library

Library of Congress Cataloging-in-Publication Data
Names: Grajales, Jacobo, author.
Title: Agrarian capitalism, war and peace in Colombia: beyond dispossession / Jacobo Grajales.
Description: Abingdon, Oxon; New York, NY: Routledge, 2021. |
Series:Routledge studies in global land and resource grabbing |
Includes bibliographical references and index. |
Identifiers: LCCN 2020056688 (print) | LCCN 2020056689 (ebook) |
ISBN 9780367469634 (hardback) | ISBN 9780367675707 (paperback) |
ISBN 9781003032236 (ebook)
Subjects: LCSH: Land tenure–Colombia–History. |
Agriculture and politics–Colombia–History. |
Agriculture and state–Colombia–History. |
Capitalism–Colombia–History. | Insurgency–Colombia–History. |
Peace-building–Colombia–History. | Colombia–History–1974-
Classification: LCC HD516.5 .G73 2021 (print) |
LCC HD516.5 (ebook) | DDC 333.3/1861–dc23
LC record available at https://lccn.loc.gov/2020056688
LC ebook record available at https://lccn.loc.gov/2020056689

ISBN: 978-0-367-46963-4 (hbk)
ISBN: 978-0-367-67570-7 (pbk)
ISBN: 978-1-003-03223-6 (ebk)

Typeset in Bembo
by Newgen Publishing UK

To Maëla and Rafaël, and in memory of Anaë

CONTENTS

ILLUSTRATIONS

Maps

Figures

Table

ACKNOWLEDGEMENTS

This book's story goes back to February 2011. I was a PhD student at Sciences Po Paris, and was invited by Jean-François Bayart to present a paper on land and violent conflict. I was absolutely thrilled, but equally terrified at the idea of sharing the floor with such brilliant minds as Nancy Peluso and Christian Lund. This meeting sent my research in a new direction. While pursuing a PhD on armed governance in Colombia, I became increasingly interested in the close-knit relationship between land and violent conflict. Nancy and Christian invited me to participate in a collective journal issue that they were putting together, and they supported me throughout the process of producing my first published paper. In the wake of this, I was introduced by Jun Borras to a vibrant and passionate community of agrarian scholars. In France, this intellectual adventure gave me the opportunity to collaborate with Jean-Pierre Chauveau and Eric Léonard, both of whom became dear friends. It even took me to Côte d'Ivoire, where doors opened at the simple mention of Jean-Pierre's name. To all my friends and mentors in this journey, I would like to express my infinite gratitude.

This was a back-and-forth movement between field research and scholarly discussion, and I know none of this would have been possible without the patience and generosity of the women and men who told me their stories of struggle, suffering, hope, and joy. None of their names are mentioned in this book, for obvious safety reasons. To them, for their confidence and support, *mil gracias*. I am particularly grateful for the support of the Colombian Commission of Jurists and its former land advocacy director Jhenifer Mojica, who provided me access to invaluable archival sources and contacts in the field. As the ideas set out in this book began to take form, they were nurtured by discussions with William Renán, Diaja Ojeda and Andrés Trujillo in Colombia, but also Romain Le Cour Grandmaison, Cécile Jouhanneau, and Marie Saiget in France. Parts of this book were presented and discussed with many friends and colleagues. Mathilde Allain, Laurent Gayer, and Benjamin Lévy had the generosity and kindness of commenting draft chapters.

Travel for field work and academic meetings, as well as the preparation of this text and the time required to write it, were made possible through financial support from various institutions. The French National Research Agency (ANR), the Institut Universitaire de France (IUF), the Foundation for the Study of Humanities (FMSH), the University of Lille, and the Centre for International Research (CERI) at Sciences Po all provided me with invaluable financial support.

I have always disliked the final part of many acknowledgement sections, where a male scholar usually thanks his spouse for her patience – which essentially lets you know who was doing the parenting and cooking during writing time. This, like countless others, is a 2020 pandemic book. I could write it at my pace because I enjoy an incredibly comfortable tenured position, contrary to many of my brilliant colleagues and friends. In addition to that, it was made possible thanks to family support, meticulous coordination, love, and solidarity – not to mention coffee. But beyond the compromises and sacrifices it required, I believe that Maëla and I have managed to live by our convictions and to keep teaching our son Rafaël that a male's achievements need not be built on men taking advantage of women's free labour. I hope that, in our family story, this book will remain a testament to this commitment.

ABBREVIATIONS

ACCU	Autodefensas Campesinas de Córdoba y Urabá – Peasant Self-Defence Groups of Córdoba and Urabá
ADM-19	Alianza Democrática M19 – Democratic Alliance M19
ADR	Agencia de Desarrollo Rural – Rural Development Agency
ANT	Agencia Nacional de Tierras – National Land Agency
ANUC	Asociación Nacional de Usuarios Campesinos – National Peasant Association
ART	Agencia de Rehabilitación del Territorio – Territorial Rehabilitation Agency
AUC	Autodefensas Unidas de Colombia – United Self-Defence Groups of Colombia
CNMH	Comisión Nacional de Memoria Histórica – National Commission for Historical Memory
CODHES	Consultoría para los Derechos Humanos y el Desplazamiento – Consultancy on Human Rights and Forced Displacement
CRIC	Consejo Regional Indígena del Cauca – Regional Indigenous Council of Cauca
CRR	Reforma Rural Integral – Comprehensive Rural Reform
DDR	Disarmament, Demobilization and Reintegration
DNP	Departamento Nacional de Planeación – National Planning Department
DOC	Departamento de Organizaciones Campesinas – Department of Peasant Organizations
ELN	Ejército de Liberación Nacional – National Liberation Army
EPL	Ejército Popular de Liberación – People's Liberation Army
FARC	Fuerzas Armadas Revolucionarias de Colombia – Revolutionary Armed Forces of Colombia

IDP	Internally Displaced Person
IGAC	Instituto Geográfico Agustín Codazzi – Colombia's cartographic and cadastral authority
Incoder	Instituto Colombiano de Desarrollo Rural – Colombian Institute of Rural Development
Incora	Instituto Colombiano de la Reforma Agraria – Colombian Institute for Agrarian Reform
JEP	Jurisdicción Especial para la Paz – Special Jurisdiction for Peace
M-19	Movimiento 19 de Abril – 19th of April Movement
OSP	Ordenamiento Social de la Propiedad – Social Property Planning
PPTP	Proyecto de Protección de Tierras y Patrimonio para la Población Desplazada – Project for the Protection of the Land and Patrimony of Displaced Populations
Sintrainagro	Sindicato nacional de trabajadores de la industria agropecuaria – National union of agro-industry workers
UFCO	United Fruit Company
UNDP	United Nations Development Programme
UP	Unión Patriótica – Patriotic Union
URT	Unidad de Restitución de Tierras – Land Restitution Unit
USAID	United States Agency for International Development
ZRC	Zona de Reserva Campesina – Peasant Reserve Zone

INTRODUCTION

Histories of land and people

A land of sorrow

I met Alberto for the first time in 2009, during the first research trip of my career.[1] He lived in the city of Santa Marta, a busy port on the Caribbean coast of Colombia. Alberto's house was not located in the colonial city centre, where backpacker hostels were beginning to multiply, nor in the showy waterfront neighbourhood of El Rodadero, where the well-off Colombians preferred to sojourn. He lived a couple of kilometres further, in Los Fundadores, a sprawling slum of 10,000 inhabitants, replete with dusty roads and shanty houses, where many exiles from the war had landed. Santa Marta was not merely a prosperous port and a touristic spot; it was also home to nearly 100,000 internal refugees, the *desplazados*, as they are called in Colombia, most of them peasants from the inner country or the coastal mountains.

Alberto was one of those *desplazados*, called IDPs (internally displaced persons) in the jargon of humanitarian professionals. He had fled his village in 2002 when a right-wing paramilitary militia killed and tortured some of his friends and neighbours after accusing them of collaborating with the rebel guerrilla group ELN (National Liberation Army). Alberto considered himself a lucky man. He believed he had probably been marked to be one of the victims, but for reasons he had never come to understand, his life had been spared. The afternoon I met him, Alberto told me the chilling story of his exile. The day the paramilitary soldiers arrived in his village, they gathered all the men and told them to go to the cassava field that was jointly farmed by the community. It was a part of a rural development project funded by foreign donors and implemented with the support of the FAO (UN Food and Agriculture Organization). They were ordered to uproot the cassava crop and fill up a trailer attached to a tractor that had been donated by an NGO a few weeks earlier. When they had finished, five men were forced to lay down on the dirt road. One of the paramilitary soldiers climbed into the tractor and ran over their bodies.

The reasons for such egregious brutality are deeply rooted in the political economy of the Colombian war. Alberto and his friends' lives epitomized many of the challenges that Colombian peasants have faced when struggling to control the land they farm. Alberto's first experiences as an activist date back to the 1970s, when the National Peasant Association (ANUC) was occupying the land of rich – and often absentee – landlords in hopes of forcing the state to fulfil the overdue promises of agrarian reform. Alberto's home back then was a village in the municipality[2] of Pivijay, located in the centre of the department[3] of Magdalena. This is a territory of extensive cattle ranching and extreme inequality, where just a few families own the lion's share of the land. Peasant mobilizations have been suppressed with brutality, as the police have taken the side of the landowners – the richest of whom have gone as far as sponsoring their own private armies staffed with local thugs who killed, tortured, and harassed prominent peasant leaders. By the late 1990s, there was little doubt that landlords had emerged victorious in these land conflicts and national government officials had ultimately lost any interest in peasant struggles or peasant agriculture, captivated as they were by the agribusiness pathway to development. However, some local inhabitants who had struggled for three decades over small plots of land were still seen as a nuisance by the most regressive of the local elites. This is when the paramilitary militias arrived in central Magdalena. Some wealthy locals welcomed them as the ones who would defend their lives and property against the combined threat of leftist guerrillas and peasant mobilizations. In the minds of both the militias and many landowners, peasants and guerrillas were conflated in a single threat to the proprietarian social order. As some members of local patrician families took the plunge and became paramilitary leaders themselves, the anti-peasant bias of the militias was further reinforced. The brutality of the day Alberto fled his land – when five men were run over by a tractor loaded with the cassava they themselves had grown and farmed – was the culmination of a story of class struggle.

The argument

The story of Alberto and his comrades illustrates two key features of the Colombian countryside. First, Colombia has one of the most unequal rural land property structures in the world. According to the latest surveys available, just 0.2 per cent of producers own estates of more than 1000 hectares, which cover a total of 32.8 per cent of the country's farmland. Conversely, 69.5 per cent of producers occupy plots of 5 hectares or less. Their properties cover only 5.2 per cent of the country's available farmland (DANE 2016). Most of the rural population lives in conditions of extreme poverty and hardship. Only 36 per cent of rural families have access to land; of these, 66.9 per cent farm plots that are too small to generate sufficient income for their material survival. Ancillary income, and in particular money sent by family members who have moved to large cities or abroad, is the only way to fulfil the most basic needs. Consequently, about half of rural households live below the poverty line, while less than a quarter of urban dwellers live in a comparable state of poverty.

Second, Colombia has been home to a long and deadly civil war, with the victim toll estimated at more than 160,000 deaths and 6.5 million forced displacements between the early 1980s and the 2010s (CNMH 2015, 2016) and violence massively concentrated in the countryside. While communist guerrillas first appeared in the 1960s and 1970s, it was in the late 1970s that clashes with them began to smack of insurrection (Echandía 2006a; Pécaut 2006). In the early 1980s, paramilitary groups became an essential part of the counterinsurgency apparatus, and, more broadly, part of the various forms of political and social repression at play (Grajales 2017; Gutiérrez 2019).

Much has been written about the articulation between these two aspects of Colombian society. For Alfredo Molano (2015) or Darío Fajardo (2014), land inequality nurtured the process of violence escalation by marginalizing recently colonized territories where communist guerrillas were particularly influential, feeding the frustrations of landless peasants and frontier settlers and generating brutal class relations in the places most directly connected to global value chains. For Alejandro Reyes (2009), the creation and development of paramilitary groups should be understood as a response from the landed elite and drug-trafficking *nouveaux riches* to the threats represented by peasant organizations and leftist parties. Abundant English-language literature dealing with the political economy of war in Colombia has concluded that there is a direct relationship between violence and accumulation, contending that war has essentially been a mechanism of capitalist accumulation and exploitation.[4] While the straightforward character of this interpretation is certainly attractive for anyone sympathetic to the struggle of impoverished peasants like Alberto, I argue that its simplicity leads many authors to miss an important point. The state, in Colombia as elsewhere, is not a unified and coherent entity, but a field of competition (Joseph and Nugent 1994; Bayart 2009; Lund 2016). Capitalism, while an overarching and encompassing economic system, does not operate in a coherent manner, but through contradictions and incongruity (Polanyi 1944; Meiksins Wood 1995; Ferguson 2010).

This book seeks precisely to address this incongruity. It asks how both war and peacemaking reshape access to land and transform rural societies, class relations, and, more generally, agrarian capitalism. With these questions in mind, I will advance two central arguments.

First, armed violence should not be analysed as an exceptional form of action. It is embedded in relations of exploitation and accumulation. It is one of a variety of possible manifestations (albeit the most extreme) of the power relations that sustain unequal access to resources. It ensues that violence should not be analysed exclusively in terms of its direct outcome (such as murder, forced displacement, and land dispossession) but more fundamentally as a factor that alters the social and political conditions of accumulation (Richards 2005a; Hoffman 2007; Debos 2016). This does not correspond to most people's views on violence. We often see armed conflict as a radical rupture in the normal course of society. But this is not simply a commonplace view among individuals. It is also firmly entrenched in the work of peacebuilding professionals, whether they work for international institutions, NGOs, or national agencies worldwide. Peacebuilding activities are grounded in the idea that a transition from war to peace amounts to a return to normality. Consequently, they often depict violence as an extraordinary form of action that is entirely divorced from social life.

This leads me to my second argument, which is that common representations of violence result in a clear differentiation between wartime plundering and legitimate accumulation (Grajales 2020a, 2020b). As such, when peacebuilders venture into the realm of the economy, they see their mission as limited to the former, while the latter should remain governed by the free market. Post-conflict policies might provide incentives for business investment or reform regulations to make them more flexible, but they do not perceive capitalist development as a potential vector of further exclusion and dispossession. This essentially ignores the violent and very recent roots of capitalism. It amounts to pretending that relations of power produced and reproduced through violence can be magically laundered whenever inequalities are no longer imposed through the use of the rifle, but rather sustained by the title.

In short, despite the fact that a fine-grained analysis of the political economy of war demonstrates that armed violence is deeply embedded in other, less extraordinary forms of material exploitation and domination, peacebuilding policies reinstate a very restrictive definition of the economies of war and peace. By the same token, they conceal forms of accumulation and exclusion that do not correspond to outright spoliation and plunder. As such, both war and peacemaking participate in the formation of an exclusionary form of agrarian capitalism. They should be studied together, as two sides of a same coin.

Land, violence, and capitalism

These questions will be addressed through an in-depth analysis of the Colombian case. Colombia offers what John Gerring calls a 'pathway case', i.e., a 'case study in which the purpose of an intensive analysis of an individual case is to elucidate causal mechanisms' (Gerring 2007). Indeed, it features the combination of two traits that are seldom found together: while it has been the location of one of the deadliest wars in contemporary Latin American history, during this period of conflict, it has possessed a relatively stable state apparatus, which resulted in an intense policymaking response to violence and its most salient consequences (Bolívar 2003; Gonzáles *et al.* 2003; Echandía 2006b). Colombia was never a 'failed state'; violence prompted the development of domestic agencies tasked with peacemaking, assistance to victims, and conflict monitoring. As contradictory as it may seem, violence did not threaten the state, but instead contributed to its continued formation (Grajales 2017; Ballvé 2020). In addition, these apparently contradictory trends took place against the backdrop of a globalizing agrarian economy marked by the development of agro-commodities (Marin-Burgos and Clancy 2017), the financialization of agriculture (Genoud 2020), and government efforts to attract and secure foreign investment (Maher 2018).

As a whole, this book is not written with the aim of documenting the Colombian case in itself, but with the aim of using it to address more general questions about the political economy of war and peace. Through this enterprise, I will engage in conversations with two main bodies of literature: those on the political economy of war and those providing a critical analysis of economic peacebuilding.

Land and the political economy of war

In Africa, South Asia, and Latin America, land tensions are arguably a primary factor nurturing instability and violence and requiring the intervention of peacemakers, development professionals, and governance experts. Over the past two decades, scholars and practitioners have increasingly come to recognize the importance of land conflicts in the emergence of armed violence and civil wars.[5] Most empirically grounded literature has moved away from popular Malthusian views – still influential among development donors – predicting that armed conflict would automatically arise from land scarcity and/or property concentration.[6] A more nuanced view examines the ways in which land becomes a vector of armed mobilization and a critical ingredient in political polarization[7] and highlights that conflicts are not solely related to the use of land, but are – more fundamentally – tied to the legitimacy and authority of the institutions regulating access to it (Peters 2013; Lund 2016).

Besides the interest for the place of land in armed conflict, a parallel research question deals with the consequences of violence on the ways in which access to land is gained and maintained (Hall *et al.* 2011; Lund 2018). However, much less literature has been produced to address this question. Specialists of civil wars have largely overlooked the contribution of violent mobilizations to economic domination and class formation.[8] Even specialists of the political economy of war, who should be much better equipped to deal with this subject, have been astonishingly shy when dealing with transformations affecting access to land in times of war (with the notable exception of Cramer 2006). When dealing with problems of capital accumulation and class formation, the emphasis has often been placed on macroeconomic problems (Ottaway 2002; Suhrke and Buckmaster 2006; Cooper and Turner 2013), corruption (Goodhand 2008; Le Billon 2008; Zaum and Cheng 2011), or high-value resources (Le Billon 2012).

Fortunately, scholars of both conflict studies and political economy have forged useful tools for the analysis of this research question. Throughout this book, I will draw on two fundamental contributions to these fields.

First, I utilize a conceptual framework that analyses war and peace as not radically distinct and often overlapping social configurations (Marchal and Messiant 1997; Richards 2005b; Cramer 2006; Debos 2016). One key consequence of this is that we should not presume a fundamental difference between wartime and peacetime accumulation, legal and illegal taking, and plunder and business (Roitman 2005; Briquet and Favarel-Garrigues 2010; Marchal 2010; Kurtenbach and Rettberg 2018). This, of course, differs from the fundamental representation that policymakers, international organizations, and development professionals promote when discussing the ties between peace, war, and the economy. But more fundamentally, it calls for a reappraisal of the place of war and violence in the transformations of contemporary capitalism (Polanyi 1944; Meiksins Wood 2002; Cramer 2006). As a result, the analysis should move from a focus on the most spectacular aspects of the wartime economy to the multiple ways in which violence fashions the political economy as a whole, i.e., the forms of accumulation, the institutions, the power practices, and the political identities.

The second idea that will prove useful in this book deals precisely with the way in which political and economic power can be integrated into a single, unified conceptual framework. I will utilize the concept of 'inequality regimes' to designate the entanglement of positions of political power and positions of accumulation, which enables powerful economic actors to shape the action of the state to protect their advantages and avoid government scrutiny (Boyer 2016). This notion stresses the fact that economic inequality, large asymmetries in the distribution of political power, and the discourse justifying both of these realities are intrinsically related and mutually reinforcing phenomena. As such, inequality regimes are characterized by the articulation between a political regime – the rules that determine the exercise of political power – and a property regime – the rules that determine how people interact with possessions and natural resources; to a certain extent, this articulation is rendered possible by a set of ideological justifications for inequalities stemming from this articulation between politics and property (Piketty 2020).

In applying these two sets of ideas, my interest throughout this book will be on the multiple ways in which armed violence interacts with pre-existing rules, institutions, and political forces, moulding an inequality regime that cannot be reduced to the dichotomy between wartime and peacetime economies.

A critique of economic peacebuilding

A further reason why social scientists should not take the divisions between war and peace for granted is that peace is an ambiguous word (Mac Ginty 2016). While every person who has experienced war has a very clear idea of what peace should look like (Courtheyn 2018), the peace we talk about in comfortable offices in Paris, London, or even Bogota has less to do with the absence of war than with a wide range of policies, narratives, and instruments promoted by peace professionals as a way out of war. 'Post-conflict transition' is less about transforming a society than about building a narrative that will garner enough support to become real.

The blurriness of the notion of peace is one of the subjects explored by a second body of literature that I hope this book can contribute to. Over the last two decades, numerous scholars have endeavoured to provide a critical evaluation of the relationship between economic policymaking and post-conflict transition.

Most of this literature originated from a critique of what came to be called the 'liberal peace' paradigm, i.e., the view that peacebuilding interventions should be concerned with creating the bedrock for liberal democracy and a free-market economy (Mac Ginty 2006; Richmond 2012). In the field of political economy, critical thinkers have considered that 'contemporary liberal thinking is erroneously idealistic about the relationship between violence and development', as it considers 'that peace ushers forth an era of peace dividends' (Selby 2008, 24–5). The collective volume put together by Pugh, Cooper, and Turner (2008) marked a milestone for this line of inquiry, compelling us to seriously consider how peacemaking can foster exclusionary dynamics and how the reproduction of inequality is concealed under the cloak of peace. More recently, the contributors gathered by Berdal and Zaum (2017) have engaged in a similar enterprise, demonstrating that post-war societies are not the 'clean slates' that peacemaking professionals tend to make them out to be, and that they are characterized not by radical socioeconomic change, but by powerful actors' capacity to consolidate their wealth and power and stall social change. Methodologically, most of this research shares the conviction – greatly influential for my work – that the study of post-conflict economic processes should pay greater attention to everyday processes of accumulation (Jennings and Bøås 2015; Distler *et al.* 2018; Jennings 2018). Overall, this literature consistently makes the case that inequalities that were produced or aggravated through violent means tend to reproduce themselves through more peaceful, institutionalized means. As these struggles often take place against the backdrop of a free-market imperative (Ottaway 2002), already dominant actors hold a distinct advantage that allows them to continue to thrive.

This book engages a dialogue with this literature in two distinct manners. First, I reflect on the shifting relationship between redistribution and peace. Post-conflict economic policies pursue economic growth in the name of peace. As it is, the delivery of jobs and material prosperity is supposed to nurture stability, with people doing business instead of waging war. Of course, the issue with this view is that it operates under the fundamentally flawed assumption that 'a radical break with the past is possible' (Torjesen 2017, 52). In reality, any transformative ambitions that governments and donors might harbour are hindered by the very policies that are part and parcel of the post-conflict toolkit. Cheng (2017) mentions three of these sources of contradiction: the priority granted to stability over most other objectives, the imperative of economic liberalization, and the support provided to state structures that formerly nurtured plunder and dispossession. Combined, these imperatives tend to favour already well-connected economic actors, who are able to use post-conflict resources to thrive as they convert their ill-acquired assets into clean capital (see, for instance, Aspinall 2009; Ballvé 2020; Grajales 2020b, 2020c).

Furthermore, when they do tackle the material consequences of war, post-conflict policies generally adopt a restorative approach, considering that the demands of internal and international refugees, as well as the global influence of the 'transitional justice' paradigm, justify a policy of asset restitution (Arbour 2007; Lai 2020). However, this policy not only generally fails to take into account the overarching effects of war, it also fundamentally requires the direct effects of war to be differentiated from other forms of material injustice (García Arboleda 2019). By the same token, it naturalizes the existing inequality regime, including its most dreadful aspects, and shuts the door on redistributive approaches to peacemaking. Together, these approaches to the post-conflict economy, which are the backbone of both domestic and international peacebuilders' intervention, tend to foster the consolidation of warless capitalism.

The case of Colombia

The characters

This book has three central characters. First, the state, which cannot be reduced to a simple tool in the hands of dominant classes, and can act as a battlefield between rival visions of agrarian societies and economies. But this battlefield, I will show, has never guaranteed competition on equal terms between these opposing visions. Not only is the influence of landed elites within the administrative and political apparatus far too entrenched, but the commitment to a more equal society among liberal sectors of the dominant classes has consistently been outranked by concerns about stability and capitalist development. Even more importantly, the well-entrenched liberal belief in the existence of an independent economic playing field whose autonomy from politics should be safeguarded – as well as in the capacity of 'the market' to generate an optimal state of social equilibrium – has been a tremendous obstacle for any progressive social transformation.

The second character comes in the form of paramilitary militias. These armed groups have been directly responsible for 59 per cent of mass-murder cases in Colombia (CNMH 2016), and in every place they held sway, there were consistently massive levels of forced displacement and land concentration (CNMH 2015). Scholarly, journalistic, and activist research has shown that they played a key role in land grabbing and dispossession, especially in regions characterized by a long history of land conflicts (Vásquez and Barrera 2018, 62–113). In doing so, they also succeeded in accumulating land for themselves or on behalf of their corporate allies (Reyes 2009; Grajales 2011a, 2013; Gutiérrez and Vargas 2016; Ballvé 2020). An impressive amount of research has now demonstrated that stories such as Alberto's, which opened this introduction, were typical of the fate of hundreds of thousands of Colombian peasants (Ibáñez and Vélez 2008; Ibáñez and Moya 2010; Gutiérrez and Vargas 2016).

The final character comes in the form of these selfsame peasants. They are analysed as a social class, i.e., a heterogeneous group whose political coherence stems from processes of social mobilization and from the characteristics of agrarian economies (Moyo and Yeros 2007). This class is not an archaic remnant of a rural past, but a group of people embedded in the agrarian capitalist system as providers of labour and food (Bernstein and Byres 2001; Akram-Lodhi and Kay 2008; Bernstein 2009; Roa-Clavijo 2021). Contrary to the romantic image proffered by some intellectuals and NGOs, most of these peasants do not aspire to achieve a state of autarky, but instead seek only to obtain a favourable position within agrarian value chains (Li 2014). The peasantry will also be analysed as a target of policy action. From agrarian reform to the promises of the Green Revolution; from poverty alleviation to the humanitarian attention paid to forcibly displaced peasants; from land restitution to the most recent perspectives of rural development and peacebuilding, the peasantry has been a subject of varying political concern: a class to control, but also to refashion, both materially and morally. A subject of 'improvement' (Li 2007).

Interactions between these three categories of actors will be analysed through a territorial and historical lens. I will deal both with the production of public policies, at the intersection of domestic and international institutions, and the struggles for land observed on a local scale – at times adopting a microview. The cornerstone of my demonstration will be provided by the joint analysis of the dynamics of violence and attempts at building peace.

Territories of violence

When I met Alberto, the paramilitary militias that had terrified him and his community were officially a thing of the past. The national confederation of the AUC (United Self-defence Groups of Colombia), along with other independent groups, had negotiated a demobilization agreement with the government. More than 30,000 men and women supposedly laid down their weapons and entered a DDR (disarmament, demobilization, and reintegration) process sponsored by foreign donors and internationally showcased as a success (Grajales 2011b; Carranza-Franco 2019). In reality, further research has shown that many of the demobilized combatants had been engaged by the paramilitary bosses to balloon the numbers, while a high percentage of well-trained militias held on to their best weaponry and joined the ranks of repurposed gang-style armed groups. Today, they generally keep a lower profile than their predecessors, but they remain key players in the cocaine-smuggling business and a primary cause of insecurity and death both in the countryside and in cities.

Where did these paramilitary militias come from? The first paramilitary groups emerged in the early 1980s at the crossroads of three related processes. First, they were rooted in histories of rural conflicts between landlords and peasants, many of them dating back to the 1960s (Machado and Meertens 2010). Second, they were also a reaction to more recent forms of social mobilization, as organized protests promoted by the left, grassroots activism, and labour movements were challenging the local order throughout Colombia (Romero 2003). Moreover, these patterns of social conflict were refashioned by Colombia's emergence as a part of global drug value chains. From the early 1970s, the subsequent marijuana and cocaine booms favoured the massive purchase of rural estates by the new narco-bourgeoisie (Reyes 1997). Landed investment led to political proximity between drug smugglers and local dominant classes. Challenges to the latter's political supremacy were seen as a threat by the former. The narcos' private armies thus became a conservative force committed to the reproduction of an existing exclusionary social order (Gutiérrez 2019). Lastly, this violent reaction was made possible by transformations in the Colombian military, which consistently engaged in a policy of dirty war as early as the late 1970s (Vásquez and Barrera 2018). Together, the radicalization of local politics, the politicization of drug barons, and the criminalization of law enforcement combined to create the first paramilitary militias (Medina Gallego 1990; Gutiérrez and Barón 2005; Ronderos 2014; Grajales 2017). Most of these local groups joined the national confederacy of the AUC, created in 1997 as a loose association of competing interests. These were the men who ravaged Alberto's community.

While the paramilitaries were not a simple reaction to guerrilla activity, support was brought to their cause by fear and rejection of the rebels' actions. Colombia holds the dubious distinction of having the oldest leftist guerrillas in the western hemisphere: the FARC (Revolutionary Armed Forces of Colombia) and the ELN (National Liberation Army). The FARC negotiated a peace settlement with the government in 2016, which is currently at the centre of much political debate on the prospects of 'post-conflict transition'. Today, while most of its leaders and militants remain committed to the peace process, a few thousand combatants belong to 'dissident' groups that refused to lay down their arms and have continued to recruit new fighters (Indepaz 2020). While the ELN never possessed the same firepower and territorial footprint as the FARC, they have taken advantage of the FARC's transition to civilian life to expand their influence and capture new sources of income from drug trafficking, oil smuggling, and racketeering.

Historically, the guerrillas emerged in marginal areas of the country: the Amazon Andean foothills and the deep Caribbean hinterland, or certain inter-Andean valleys and border zones (Aguilera Peña 2014). Very often, their growth was associated with movements of internal migration, as landless peasants colonized the country's agrarian frontier. Their development was enabled not by statelessness, but, more accurately, by state marginalization, as people and territories were pushed back beyond the perimeter of the imagined nation-state (Jaramillo *et al.* 1986; Gonzáles *et al.* 2003). Along with the social grounding of rebel groups, these conditions of marginality determined the geography of coca production, which massively sprung up in areas of failed development, where the promises of the frontier had been frustrated by a lack of government support that had made it unrealistic for conventional agriculture to prosper (Torres 2018). The guerrillas, in turn, established a social base by regulating the relationships between peasant coca producers and the intermediaries in the cocaine value chain, before themselves going on to become key actors in the cocaine business (Molano 2010; Ramírez 2011).

Conversely, paramilitary groups emerged in more central zones characterized by extensive cattle breeding and export-crop production (Gutiérrez 2019). In many cases, these were areas of entrenched social conflicts where land had been massively accumulated by landlords and agribusiness entrepreneurs. These were the social groups that were threatened by the popular mobilizations of the early 1980s in the cities and the countryside. They were also those who had the most to lose from the guerrillas' expansion from their strongholds on the edges of the country towards its centre. Beginning in the late 1970s, the FARC and the ELN, not to mention the guerrillas from the EPL (People's Liberation Army) and M-19 (19th of April Movement), were progressively gaining the capacity to strike at the core of the nation's institutions and economy. By the 1990s, territorial competition had begun to feed an unforeseen escalation in the levels of violence. From their strongholds, paramilitary militia groups would expand to the periphery of capitalist development, where guerrilla groups (especially the FARC) were the prevailing political authorities (Gonzáles *et al.* 2003). While necessarily simplistic, the dynamics of the armed conflicts of the 1990s can be boiled down to this twofold expansion, with guerrillas moving inward from the outskirts to the centre, and paramilitaries expanding from the centre into the outskirts (Echandía 2006b; Vásquez and Barrera 2018).

A transition to what?

While violence levels remain high and mass murders have returned to newspaper headlines in 2020, very few people would classify current-day Colombia as a country at war. After all, this is an internationally saluted case of peacemaking, with the state asserting its commitment to imposing 'peace with legality' over territories held for decades by insurgent and paramilitary groups. In reality, this book will also show that Colombia is yet another case in which the word 'transition' is highly misleading. Scholars such as Paul Richards (2005b) and Marielle Debos (2016) have demonstrated that contemporary civil wars seldom result in clear-cut peace. While they are distinguishable through the transformations that have occurred in the way violence is used and in the expectations of ordinary people, as well as the more or less coherent implementation of the political engineering of peacebuilding that now accompanies them, they are first and foremost uncertain times. I use the word 'warless' to address this uncertainty. This is no longer war, but it most certainly does not qualify as peace.

The concern with the idea of 'transition' is not a recent one in Colombia. For decades, peacebuilding has been a recurring practice and a source of much policy-making activity. As early as the 1980s, various cycles of negotiation between the government and the guerrillas were attempted (Chernick 2008; Rettberg 2012). They were not entirely fruitless, as they resulted in separate peace settlements with the M-19, the EPL, and other minor groups between 1990 and 1994. In those same years, negotiations were also held with paramilitary militias, although in a much more discreet fashion, and they resulted in the very partial demobilization of a group commanded by militia boss Fidel Castaño on the southern Caribbean coast (Córdoba and Urabá).

The escalation of war, US pressure on Colombia to deal with paramilitary militias as ordinary drug traffickers, and favourable prospects offered by Álvaro Uribe's hard-line right-wing administration explain the paramilitaries' eagerness to negotiate in 2003. Most groups officially laid down their arms between 2003 and 2006, with the aforementioned mixed results. However, while DDR was not the end of paramilitarism, it did reveal the underlying mechanisms of paramilitary rule. One of the aspects that was revealed concerned land. As access to their former strongholds became easier, and paramilitary bosses themselves began to tell their side of the story, judicial, scholarly, and journalistic research unveiled the massive land grabbing that had generally been undertaken by these militias (Salinas Abdala and Zarama Santacruz 2012). Tales of people like Alberto, of massive dispossession and exile, became common in newspapers and research reports. They were complemented by a finer understanding of the bureaucratic mechanisms used to launder dispossessed land and to transform it into a tradable asset.[9] The insight that was gained into these forms of dispossession contributed to putting violent land grabbing on the political agenda and making land restitution a foreseeable perspective. However important this attention was, it had the downside of transforming dispossession into a self-standing reality, the result of massive and extraordinary violence that was somehow too extreme to be part of the normal functioning of society. Such a view tended to disconnect land dispossession from its political and economic substrate, concealing its embeddedness in the historical transformations of agrarian capitalism.

In September 2012, the Colombian government and the FARC announced their will to reach a negotiated settlement to the armed conflict. An agenda for the peace talks was made public, with the first chapter corresponding to agrarian policies (García Trujillo 2020). To a certain extent, the negotiations that took place in the city of Havana between 2012 and 2016 brought with them the promise of connecting the treatment of egregious forms of violence like those that were suffered by Alberto and his community to a transformation of the larger inequality regime. Since the September 2016 signature of the peace agreement, which resulted in the demobilization of most of the FARC combatants and the transformation of the movement into a political party, most of these promises have been frustrated. One of the objectives of this book is to tell the story of this failure, not only through a careful account of its unfolding,[10] but also by connecting it to a longer history of war and peacemaking.

What will follow

My study is based on extensive field research conducted between 2009 and 2019 in several areas of Colombia. I will also use statistical data from public sources and satellite imaging data. Historical sections – in particular in Chapters 1 and 4 – rely heavily on the works of historians, sociologists, economists, and political scientists, as I aim to present a comprehensive view of the existing literature. My research was centred around two lines of inquiry. First, I worked on the articulation between land access and violence. The core of this first line of research is provided by a case study on the agribusiness zone of the Magdalena department, a coastal territory on the northern Caribbean coast. Second, the book deals with the multiple interconnections between peacebuilding and land through an analysis of public policies and peacebuilding efforts. This second line of inquiry draws on dozens of interviews with policymakers, development professionals, and NGO staff. These sources are complemented by a systematic review of the extant literature, as well as the consultation of numerous policy documents. In all, I conducted nearly 200 semi-structured interviews with a great variety of people, ranging from peasants and internal exiles to government officials and foreign development experts. While I do not aspire to provide a full account of all that I learned from those I have spoken with over the course of these years, I certainly do hope that the pages that will follow will never betray their confidence.

Chapter 1 traces the historical formation of a land inequality regime. It ties the structure of landed property in Colombia to the conditions of the inception of agrarian capitalism from the end of the 19th century to the expansion of the agrarian frontier. It traces the struggles of peasants across the 20th century and examines the failed history of agrarian reform.

Chapters 2 and 3 move to a local scale of analysis, and are focused on paramilitary militias. Chapter 2 provides a historical analysis of how war unfolded in the territory of Magdalena. It studies paramilitary violence as intrinsically associated with the reproduction of an inequality regime. Chapter 3 adopts an even more microperspective. Through the analysis of a set of labour and land conflicts that occurred in the 1990s and early 2000s in the agribusiness zone of Magdalena, it shows how land access and dispossession is determined both by economic domination and the capacity of powerful actors to use violence against those who challenge their position of dominance.

Chapter 4 moves back to the national and international scales. It deals with the emergence of IDPs – people like Alberto – as subjects of rights. It shows how the development of a humanitarian attention policy refashioned the agrarian issue. With one of the largest populations of IDPs in the world – most of them rural dwellers – Colombian institutions in the early 2000s were also facing a problem of massive land dispossession. This chapter describes the transformations of this political problem and the way in which it was addressed.

Chapter 5 reflects on the failed attempt to open up the policy agenda to a larger discussion on redistribution. During the peace talks between the government and the FARC in Havana, the issue of land ranked high on the list of priorities. However, the ambitions of the agreement were rapidly frustrated, as the country saw a deterioration of the political climate, the administration's incapacity to move towards effective implementation, and conservative backlash from reactionary sectors of society – many of them members of the landed elites.

Chapter 6 returns to a local scale to reflect on how the most recent transformations in the Colombian conflict translate in a territory widely considered as 'pacified'. Magdalena today is widely seen by political decision-makers and commentators as having left behind its history of violent paramilitary rule. It is one of those regions in which a modern and emerging Colombia is supposedly being shaped. A closer look shows how an exclusionary social order, previously consolidated through egregiously brutal means, no longer needs to rely on overt and massive violence.

Lastly, a conclusion brings the analysis to a close and highlights the relationships between the different parts of the book. It then analyses the central contributions of this text to the study of the political economy of war and peacemaking. The final pages attempt to critically evaluate the ways in which the development and peacebuilding communities analyse the role of land in societies supposedly transitioning from war to peace. I then venture to draw some general conclusions from the shortcomings of the Colombian case that can be useful not only for academics, but also for all those interested in dealing with intractable and violent struggles for land.

As I write the last words of this book, the prospects of durable peace in Colombia are murkier than ever. Murders of activists and social leaders have become widespread. Depending on the methodology and the manner in which victims are classified, estimates about the number of activists murdered between November 2016 and mid-2020 vary from 500 to nearly a thousand. Most of them were men and women like Alberto. Peasant leaders, fighting to acquire ownership of their lands and to protect the environment and their livelihoods. These murders have many causes, and this book is not an inquiry into this new breed of violence. However, the message that I expect it to convey is that an economic system built through war will not be magically turned into a pacified open market when the war is over. As long as the violent roots of this exclusionary social order are not addressed, its beneficiaries and guardians will not hesitate to return to the same brutal methods, time and time again.

Notes

1 The names of interviewees have been fictionalized so as to protect their identities.
2 Municipalities (*municipios*) in Colombia, as in many other Spanish-speaking countries, are the smallest administrative unit. They are similar to counties and generally comprise a municipal seat (the main town), smaller towns, and rural areas.
3 In the administrative divisions of Colombia, departments (*departamentos*) are the intermediate level of government, between the central state and the municipalities.
4 This premise is shared by works as diverse as those of Avilés (2007), Hristov (2014), and Maher (2018). These authors intend to single out the intimate link between violence and capitalism in Colombia. Hristov, in particular, considers that paramilitary groups were part and parcel of a larger apparatus of accumulation, which includes agribusiness conglomerates and the state.
5 See, among many others: Huggins (2005), Baranyi and Weitzner (2006), Peters and Richards (2011), Bavinck *et al.* (2014), Cuvelier *et al.* (2014), Van Leeuwen and Van Der Haar (2016), and Klaus (2020).
6 The work of Homer-Dixon (1999), widely criticized but still influential, is an example of this kind.
7 An argument made, among others, by Le Billon (2001), Peluso and Watts (2001), Cramer and Richards (2011), Boone (2014), and Chauveau *et al.* (2020).
8 This is the case with Kalyvas (2006) and much literature inspired by his work. In the case of Colombia, see Arjona (2017).

9 An abundant amount of literature on dispossession was produced from the late 2000. For a thorough presentation, see Gutiérrez and Vargas (2016).
10 An objective already fulfilled by García Trujillo's (2020) important book.

References

Aguilera Peña, M. 2014. *Guerrilla y población civil: trayectoria de las FARC 1949–2013*. Bogotá: Centro Nacional de Memoria Histórica.

Akram-Lodhi, A.H. and C. Kay, eds. 2008. *Peasants and Globalization: Political Economy, Rural Transformation and the Agrarian Question*. London: Routledge.

Arbour, L. 2007. Economic and Social Justice for Societies in Transition. *New York University Journal of International Law and Politics*, 40(1), 1–27.

Arjona, A. 2017. *Rebelocracy: Social Order in the Colombian Civil War*. New York, NY: Cambridge University Press.

Aspinall, E. 2009. Combatants to Contractors: The Political Economy of Peace in Aceh. *Indonesia*, (87), 1–34.

Avilés, W. 2007. *Global Capitalism, Democracy, and Civil-Military Relations in Colombia*. Albany, NY: SUNY Press.

Ballvé, T. 2020. *The Frontier Effect: State Formation and Violence in Colombia*. Ithaca, NY: Cornell University Press.

Baranyi, S. and V. Weitzner. 2006. *Transforming Land-Related Conflict: Policy, Practice and Possibilities*. Ottawa: The North-South Institute.

Bavinck, M., L. Pellegrini, and E. Mostert, eds. 2014. *Conflicts over Natural Resources in the Global South: Conceptual Approaches*. Londres: CRC Press.

Bayart, J.-F. 2009. *The State in Africa*. Cambridge: Polity.

Berdal, M. and D. Zaum, eds. 2017. *Political Economy of Statebuilding: Power after Peace*. London: Routledge.

Bernstein, H. 2009. 'The Peasantry' in Global Capitalism: Who, Where and Why? *Socialist Register*, 37, 25–51.

Bernstein, H. and T.J. Byres. 2001. From Peasant Studies to Agrarian Change. *Journal of Agrarian Change*, 1(1), 1–56.

Bolívar, I.J. 2003. *Violencia Política y Formación Del Estado*. Bogotá: Cinep.

Boone, C. 2014. *Property and Political Order in Africa: Land Rights and the Structure of Politics*. New York, NY: Cambridge University Press.

Boyer, R. 2016. A World of Contrasted but Interdependent Inequality Regimes: The Latin America Paradox. *Review of Political Economy*, 28(1), 1–22.

Briquet, J.-L. and G. Favarel-Garrigues, eds. 2010. *Organized Crime and States: The Hidden Face of Politics*. Basingstoke: Palgrave Macmillan.

Carranza-Franco, F. 2019. *Demobilisation and Reintegration in Colombia: Building State and Citizenship*. London: Routledge.

Chauveau, J.-P., J. Grajales, and E. Léonard. 2020. Foncier et violences politiques en Afrique: Pour une approche continuiste et processuelle. *Revue internationale d'études du développement*, 3(223).

Cheng, C. 2017. Private and Public Interests: Informal Actors, Informal Influence, and Economic Order after War. *In*: M. Berdal and D. Zaum, eds. *Political Economy of Statebuilding: Power after Peace*. London: Routledge, pp. 63–77.

Chernick, M.W. 2008. *Acuerdo posible: solución negociada al conflicto armado colombiano*. Bogotá: Ediciones Aurora.

CNMH (Centro Nacional de Memoria Histórica). 2015. *Una nación desplazada. Informe nacional del desplazamiento forzado en Colombia*. Bogotá: Centro Nacional de Memoria Histórica.

CNMH (Centro Nacional de Memoria Histórica). 2016. *Basta ya. Colombia: memorias de guerra y dignidad*. Bogotá: Centro Nacional de Memoria Histórica.

Cooper, N. and M. Turner. 2013. The Iron First of Liberal Intervention inside the Velvet Glove of Kantian Idealism: A Response to Burke. *Critical Studies on Security*, 1(1), 35–41.

Courtheyn, C. 2018. Territories of Peace: Alter-Territorialities in Colombia's San José de Apartadó Peace Community. *The Journal of Peasant Studies*, 45(7), 1432–59.

Cramer, C. 2006. *Civil War Is Not a Stupid Thing: Accounting for Violence in Developing Countries*. London: Hurst.

Cramer, C. and P. Richards. 2011. Violence and War in Agrarian Perspective. *Journal of Agrarian Change*, 11(3), 277–97.

Cuvelier, J., K. Vlassenroot, and N. Olin. 2014. Resources, Conflict and Governance: A Critical Review. *The Extractive Industries and Society*, 1(2), 340–50.

DANE (Departamento Administrativo Nacional de Estadística). 2016. *Tercer censo nacional agropecuario*. Bogotá: DANE.

Debos, M. 2016. *Living by the Gun in Chad: Combatants, Impunity and State Formation*. London: Zed Books.

Distler, W., E.B. Stavrevska, and B. Vogel. 2018. Economies of Peace: Economy Formation Processes and Outcomes in Conflict-Affected Societies. *Civil Wars*, 20(2), 139–50.

Echandía, C. 2006a. Conducta de la guerrilla durante el gobierno de Uribe Vélez: de las lógicas de control territorial a las lógicas de control estratégico. *Análisis político*, (57), 31–54.

Echandía, C. 2006b. *Dos décadas de escalamiento del conflicto armado en Colombia. 1986-2006*. Bogotá: Universidad Externado de Colombia.

Fajardo, D. 2014. *Las guerras de la agricultura colombiana, 1980-2010*. Bogotá: ILSA.

Ferguson, J. 2010. The Uses of Neoliberalism. *Antipode*, 41, 166–84.

García Arboleda, J.F. 2019. *El exterminio de la isla de Papayal: Etnografías sobre el Estado y la construcción de paz en Colombia*. Bogotá: Editorial Pontificia Universidad Javeriana.

García Trujillo, A. 2020. *Peace and Rural Development in Colombia. The Window for Distributive Change in Negotiated Transitions*. London: Routledge.

Genoud, C. 2020. Access to Land and the Round Table on Sustainable Palm Oil in Colombia. *Globalizations*, 18(3), 372–89.

Gerring, J. 2007. Is There a (Viable) Crucial-Case Method? *Comparative Political Studies*, 40(3), 231–53.

Gonzáles, F., I.J. Bolívar, and T. Vásquez. 2003. *Violencia política en Colombia. De la nación fragmentada a la construcción del Estado*. Bogotá: Cinep.

Goodhand, J. 2008. War, Peace, and the Places in Between. Why Borderlands Are Central? *In*: M.C. Pugh, N. Cooper, and M. Turner, eds. *Whose Peace?: Critical Perspectives on the Political Economy of Peacebuilding*. Basingstoke: Palgrave Macmillan, pp. 225–43.

Grajales, J. 2011a. The Rifle and the Title: Paramilitary Violence, Land Grab and Land Control in Colombia. *The Journal of Peasant Studies*, 38(4), 771–92.

Grajales, J. 2011b. El proceso de desmovilización de los paramilitares en Colombia: entre lo político y lo judicial. *Desafíos*, 23(2), 149–94.

Grajales, J. 2013. State Involvement, Land Grabbing and Counter-Insurgency in Colombia. *Development and Change*, 44(2), 211–32.

Grajales, J. 2017. *Gobernar en medio de la violencia. Estado y paramilitarismo en Colombia*. Bogotá: Editorial Universidad del Rosario.

Grajales, J. 2020a. Losing Land in Times of Peace: Post-War Agrarian Capitalism in Colombia and Côte d'Ivoire. *Journal of Peasant Studies*. DOI: 10.1080/03066150.2019.1691535

Grajales, J. 2020b. A Land Full of Opportunities? Agrarian Frontiers, Policy Narratives and the Political Economy of Peace in Colombia. *Third World Quarterly*, 47(7), 1141–60.

Grajales, J. 2020c. From War to Wealth? Capitalist Peace and Land Policies in Côte d'Ivoire. *Review of African Political Economy*, 47(163), 78–94.

Gutiérrez, F. 2019. *Clientelistic Warfare: Paramilitaries and the State in Colombia*. Oxford: Peter Lang.

Gutiérrez, F. and M. Barón. 2005. *Re-Stating the State: Paramilitary Territorial Control and Political Order in Colombia (1978-2004)*. Crisis States Programme DESTIN-London School of Economics, No. Working Paper no. 66.

Gutiérrez, F. and J. Vargas. 2016. *El despojo paramilitar y su variación: quiénes, cómo, por qué*. Bogotá: Universidad del Rosario.

Hall, D., P. Hirsch, and T.M. Li. 2011. *Powers of Exclusion: Land Dilemmas in Southeast Asia*. Honolulu, HI: University of Hawaii Press.

Hoffman, D. 2007. The Meaning of a Militia: Understanding the Civil Defence Forces of Sierra Leone. *African Affairs*, 106(425), 639–62.

Homer-Dixon, T.F. 1999. *Environment, Scarcity, and Violence*. Princeton, NJ: Princeton University Press.

Hristov, J. 2014. *Paramilitarism and Neoliberalism: Violent Systems of Capital Accumulation in Colombia and Beyond*. London: Pluto Press.

Huggins, C.C. 2005. *From the Ground Up: Land Rights, Conflict and Peace in Sub-Saharan Africa*. Pretoria: Institute for Security Studies.

Ibáñez, A.M. and A. Moya. 2010. Vulnerability of Victims of Civil Conflicts: Empirical Evidence for the Displaced Population in Colombia. *World Development*, 38(4), 647–63.

Ibáñez, A.M. and C.E. Vélez. 2008. Civil Conflict and Forced Migration: The Micro Determinants and Welfare Losses of Displacement in Colombia. *World Development*, 36(4), 659–76.

Indepaz. 2020. *Los grupos pos-FARC-EP: un escenario complejo*. Bogota: Indepaz.

Jaramillo, J.E., L. Mora, and F. Cubides. 1986. *Colonización, coca y guerrilla*. Bogotá: Alianza Editorial Colombiana.

Jennings, K.M. 2018. Peacekeeping as Enterprise: Transaction, Consumption, and the Political Economy of Peace and Peacekeeping. *Civil Wars*, 20(2), 238–61.

Jennings, K.M. and M. Bøås. 2015. Transactions and Interactions: Everyday Life in the Peacekeeping Economy. *Journal of Intervention and Statebuilding*, 9(3), 281–95.

Joseph, G.M. and D. Nugent. 1994. *Everyday Forms of State Formation: Revolution and the Negotiation of Rule in Modern Mexico*. Durham, NC: Duke University Press.

Kalyvas, S. 2006. *The Logic of Violence in Civil War*. New York, NY: Cambridge University Press.

Klaus, K. 2020. *Political Violence in Kenya: Land, Elections, and Claim-Making*. Cambridge: Cambridge University Press.

Kurtenbach, S. and A. Rettberg. 2018. Understanding the Relation between War Economies and Post-War Crime. *Third World Thematics: A TWQ Journal*, 3(1), 1–8.

Lai, D. 2020. *Socioeconomic Justice: International Intervention and Transition in Post-War Bosnia and Herzegovina*. Cambridge: Cambridge University Press.

Le Billon, P. 2001. The Political Ecology of War: Natural Resources and Armed Conflicts. *Political Geography*, 20(5), 561–84.

Le Billon, P. 2008. Corrupting Peace? Peacebuilding and Post-Conflict Corruption. *International Peacekeeping*, 15(3), 344–61.

Le Billon, P. 2012. *Wars of Plunder: Conflicts, Profits and the Politics of Resources*. New York, NY: Columbia University Press.

Li, T.M. 2007. *The Will to Improve: Governmentality, Development, and the Practice of Politics.* Durham, NC: Duke University Press.

Li, T.M. 2014. *Land's End: Capitalist Relations on an Indigenous Frontier.* Durham, NC: Duke University Press.

Lund, C. 2016. Rule and Rupture: State Formation Through the Production of Property and Citizenship. *Development and Change*, 47(6), 1199–228.

Lund, C. 2018. Predatory Peace. Dispossession at Aceh's Oil Palm Frontier. *The Journal of Peasant Studies*, 45(2), 431–52.

Mac Ginty, R. 2006. *No War, No Peace: The Rejuvenation of Stalled Peace Processes and Peace Accords.* Basingstoke: Palgrave Macmillan.

Mac Ginty, R. 2016. *International Peacebuilding and Local Resistance: Hybrid Forms of Peace.* Basingstoke: Palgrave Macmillan.

Machado, A. and D. Meertens, eds. 2010. *La tierra en disputa. Memorias de despojo y resistencia en la Costa Caribe (1960-2010).* Bogotá: Centro Nacional de Memoria Histórica.

Maher, D. 2018. *Civil War and Uncivil Development: Economic Globalisation and Political Violence in Colombia and Beyond.* New York, NY: Springer.

Marchal, R. 2010. Monetary Illegalism and Civil War: The Case of Somalia. *In*: J.-L. Briquet and G. Favarel-Garrigues, eds. *Organized Crime and States: The Hidden Face of Politics.* Basingstoke: Palgrave Macmillan, pp. 221–46.

Marchal, R. and C. Messiant. 1997. *Les Chemins de la guerre et de la paix. Fins de conflit en Afrique orientale et australe.* Paris: Karthala.

Marin-Burgos, V. and J.S. Clancy. 2017. Understanding the Expansion of Energy Crops beyond the Global Biofuel Boom: Evidence from Oil Palm Expansion in Colombia. *Energy, Sustainability and Society*, 7(21).

Medina Gallego, C. 1990. *Autodefensas, Paramilitares y Narcotráfico En Colombia.* Bogotá: Documentos Peridísticos.

Meiksins Wood, E. 1995. *Democracy against Capitalism: Renewing Historical Materialism.* Cambridge: Cambridge University Press.

Meiksins Wood, E. 2002. *The Origin of Capitalism: A Longer View.* London: Verso.

Molano, A. 2010. *Selva adentro. Una historia oral de la colonización del Guaviare.* Bogotá: Punto de Lectura.

Molano, A. 2015. *Fragmentos de la historia del conflicto armado (1920-2010).* Bogotá: Espacio Crítico.

Moyo, S. and P. Yeros. 2007. *Reclaiming the Nation: The Return of the National Question in Africa, Asia and Latin America.* London; New York, NY: Pluto Press.

Ottaway, M. 2002. Rebuilding State Institutions in Collapsed States. *Development and Change*, 33(5), 1001–23.

Pécaut, D. 2006. *Crónica de cuatro décadas de política colombiana.* Bogotá: Norma.

Peluso, N.L. and M. Watts, eds. 2001. *Violent Environments.* Ithaca, NY: Cornell University Press.

Peters, P.E. 2013. Land Appropriation, Surplus People and a Battle over Visions of Agrarian Futures in Africa. *The Journal of Peasant Studies*, 40(3), 537–62.

Peters, K. and P. Richards. 2011. Rebellion and Agrarian Tensions in Sierra Leone. *Journal of Agrarian Change*, 11(3), 377–95.

Peters, P.E. 2013. Land Appropriation, Surplus People and a Battle over Visions of Agrarian Futures in Africa. *The Journal of Peasant Studies*, 40(3), 537–62.

Piketty, T. 2020. *Capital and Ideology.* Cambridge: The Belknap Press.

Polanyi, K. 1944. *The Great Transformation: The Political and Economic Origins of Our Time.* New York, NY: Farrar & Rinehart.

Pugh, M.C., N. Cooper, and M. Turner. 2008. Introduction. *In*: *Whose Peace?: Critical Perspectives on the Political Economy of Peacebuilding*. Basingstoke: Palgrave Macmillan, pp. 1–11.

Ramírez, M.C. 2011. *Between the Guerrillas and the State: The Cocalero Movement, Citizenship, and Identity in the Colombian Amazon*. Durham, NC: Duke University Press.

Rettberg, A. 2012. Construcción de Paz En Colombia: Contexto y Balance. *In*: A. Rettberg, ed. *Construcción de paz en Colombia*. Bogotá: Universidad de los Andes, pp. 3–50.

Reyes, A. 1997. Compra de tierras por narcotraficantes. *In*: F. Thoumi, ed. *Drogas ilícitas en Colombia. Su impacto económico, político y social*. Bogotá: Ariel Ciencia Política, pp. 279–346.

Reyes, A. 2009. *Guerreros y campesinos. El despojo de la tierra en Colombia*. Bogotá: FESCOL/Norma.

Richards, P. 2005a. New War: An Ethnographic Approach. *In*: P. Richards, ed. *No Peace, No War: An Anthropology of Contemporary Armed Conflicts*. Oxford: James Currey, pp. 1–21.

Richards, P., ed. 2005b. *No Peace, No War: An Anthropology of Contemporary Armed Conflicts*. Oxford: James Currey.

Richmond, O. 2012. *A Post-Liberal Peace*. London: Routledge.

Roa-Clavijo, F. 2021. *Rethinking Rural Development, Food and Agriculture in Colombia*. London: Routledge.

Roitman, J. 2005. *Fiscal Disobedience: An Anthropology of Economic Regulation in Central Africa*. Princeton, NJ: Princeton University Press.

Romero, M. 2003. *Paramilitares y autodefensas. 1982–2003*. Bogotá: Planeta.

Ronderos, M.T. 2014. *Guerras recicladas. Una historia periodística del paramilitarismo en Colombia*. Bogotá: Aguilar.

Salinas Abdala, Y. and J.M. Zarama Santacruz. 2012. *Justicia y paz. Tierras y territorios en las versiones de los paramilitares*. Bogotá: Centro Nacional de Memoria Histórica.

Selby, J. 2008. The Political Economy of Peace Processes. *In*: M.C. Pugh, N. Cooper, and M. Turner, eds. *Whose Peace?: Critical Perspectives on the Political Economy of Peacebuilding*. Basingstoke: Palgrave Macmillan, pp. 13–31.

Suhrke, A. and J. Buckmaster. 2006. Aid, Growth and Peace: A Comparative Analysis. *Conflict, Security & Development*, 6(3), 337–63.

Torjesen, S. 2017. Transition from War to Peace: Stratification, Inequality and Post-War Economic Reconstruction. *In*: M. Berdal and D. Zaum, eds. *Political Economy of Statebuilding: Power After Peace*. London: Routledge, pp. 48–61.

Torres, M.-C. 2018. The Making of a Coca Frontier: The Case of Ariari, Colombia. *In*: P. Gootenberg and L.M. Davalos, eds. *The Origins of Cocaine: Colonization and Failed Development in the Amazon Andes*. London: Routledge, pp. 133–59.

Van Leeuwen, M. and G. Van Der Haar. 2016. Theorizing the Land–Violent Conflict Nexus. *World Development*, 78, 94–104.

Vásquez, T. and V.A. Barrera. 2018. *Paramilitarismo: balance de la contribución del CNMH al esclarecimiento histórico*. Bogotá: Centro Nacional de Memoria Histórica.

Zaum, D. and C. Cheng, eds. 2011. *Corruption and Peacebuilding: Selling the Peace?* London: Routledge.

1

THE POLITICAL PRODUCTION OF INEQUALITY

Inequality is not the unexpected work of the invisible hand of the market. Its underlying violence requires constant political and ideological support, mediated by the law and by the state's everyday action. A conceptualization in terms of inequality regimes, defined in the introduction as the interlocking of positions of political power and positions of accumulation, points to the importance of simultaneously embracing transformations of political power and the production of economic domination. That is the view adopted by the present chapter, which intends to reinscribe today's land access inequality in Colombia within the context of its longer history.

Centred on struggles for rural land, the following pages examine the historical production of a proprietarian social order, i.e., a 'social order based on a quasi-religious defence of property rights as the sine qua non of social and political stability' (Piketty 2020, 1044). This social order is the result of two interrelated processes. First, it stems from large landowners' and land investors' capacity to consistently turn the state and the law into an instrument for protecting their access to land. This feature is particularly important in settler states like Colombia, where property is the result of the original dispossession of Indigenous peoples and the transformation of the entire colony into a *terra nullius*. As such, the state holds the fundamental authority of transforming public land into private property. While political thinkers from at least the middle of the 19th century defended the prospects of equal access to land for the rural poor (many of whom were precisely the descendants of dispossessed Indigenous subjects and enslaved African labourers), most of these proposed distributive legal provisions remained a dead letter, as state forces were used to consolidate the monopolization of land into the hands of the few.

The second social process behind the production of a proprietarian social order is the political marginalization of peasants. From the very emergence of

MAP 1.1 Colombia with places mentioned in the book. Source: the author, created using data from Naturalearth-data.com (public domain) and OpenStreetMap (Open Database Licence – OdbL)

the peasantry as a social class in the 1920s, it became the object of contrasting representations in circles of power. For some, the activity of organized peasants needed to be controlled and put to use, so that they could be transformed into agents of modernity, while for others, their activism called for nothing more than repression. The former view was defended by 'agrarianist' political thinking and at times managed to influence policy conception and design. However, its influence was repeatedly subject to resistance from a combination of political and economic interests who felt threatened by the prospect of the peasantry becoming a political subject.

This chapter will tell the story of the triumph of proprietarian ideology and the social struggles that vigorously opposed it. The first section will show how the 'agrarian issue' emerged in Colombian politics, as both an instance of capitalist primitive accumulation and a political problem to be addressed. The second section will retrace the various attempts to implement an agrarian reform and show how these endeavours were fought by powerful proprietarian interests. The third section will reflect on the most recent period, when political liberalism ultimately worked against the interests of peasants, as class and ethnic identity were transformed into two opposing registries of social struggle.

The origins of the agrarian issue

Between the mid-19th and the mid-20th centuries, Colombia went from a rural society with a poor and scattered agrarian economy to a nation connected to global capitalism and featuring promising business opportunities (Henderson 2001). This section will retrace the major events of this century-long period. As the economy grew and the agrarian frontier expanded, a scramble for land unfolded in which settlers and landlords would compete on highly unequal terms. The monopolization of land in the frontier would give birth to the first agrarian movements, but also to a modernist elite reaction seeking to formulate class compromises in the name of both stability and economic development. The contradictions that were temporarily soothed by the Liberal Party's plans for modernization in the 1930s violently returned to the political forefront during the 1950s civil war. A reconstruction and reconciliation policy, implemented from 1958, would once again frame land as a critical problem, both for social peace and economic growth. But the failed promises of peace in the 1960s did nothing more than sow the seeds of new cycles of violence.

The struggle of the title and the axe

Over the last decades of the 19th century, Colombia, like many other peripheral nations, progressively began to assume a role in the 'first globalization' as a provider of tropical products such as quinine, cotton, tobacco, and most importantly coffee (Palacios 2002). Control over both land and labour became two interlocked challenges, as landlords and businessmen strived to master the natural resources of the frontier and the labour necessary to extract them. These attempts at exploitation gave rise to various forms of resistance that would become coordinated and politically articulated by the 1920s. The struggle between businessmen and settlers – or between the 'title and the axe', as it has been referred to in Colombian politics, finally came to an end in 1936 with the passage of a law recognizing the claims of large landowners and consolidating a proprietarian social regime.

While the Spanish crown and the early independent Republic distributed titles to huge swathes of land, most of the Colombian territory remained unoccupied and unexploited until the second half of the 19th century (Posada Carbó 1998; Román 2014; LeGrand 2016, 19–20). Colombia was a mountain and coastal society, with its populated centres concentrated in the Andean highlands where Spaniards first settled and along a few spots of the Caribbean coast, seldom connected to the country's central regions (Múnera 2008). The Caribbean hinterlands and the tropical valleys of the inner country were beyond the reach of a poor and poorly integrated economy (Safford 2010; LeGrand 2016, 15–8).

From the 1850s, the first waves of capitalist investors explored the possibilities afforded by the tropical plains and valleys beyond the old Spanish-era haciendas (LeGrand 2016, 21-sq). In the final decades of the century, the development of river navigation and the incipient but decisive tracing of roads and railroads would connect previously autarkic and parochial economies (Ocampo 1996). Export crops such as bananas and coffee (Bergquist 1986; Bucheli 2005), in addition to food for the domestic market such as meat from cattle breeding in the Caribbean plains (Ocampo 2007; Van Ausdal 2009), would open the agrarian frontier.

This posed the question of the appropriation of public lands. Even if frontier businessmen were well connected and could defend their interests before Bogota authorities, the ideal of an open frontier giving birth to a republic of independent peasants was appealing to political elites across the ideological spectrum. For Conservatives, it spoke to the moderating power of individual ownership. For Liberals, it promised a definite break with the Spanish 'feudal' past and a path to modern citizenship. For all of them, it was the promise of economic development spurred by an industrious class of settlers. In spite of the paradoxical influence of these pro-peasant ideas, this republic of independent producers remained a Utopia. Except for some coffee-growing areas of Southern Antioquia and Caldas, where family farming rose to dominance – although not without social and racial conflicts (Christie 1978; Appelbaum 2003) – land in the frontier was accumulated at a massive scale. Accumulators took advantage of local political networks, benefitted from the support of local law enforcement, and were able to defend their rights in Bogota if necessary. Accumulation often unfolded according to a two-stage process. In open frontier areas, the first settlers were independent peasants from densely populated areas who were generally seeking to escape from their dominated status as tenants. After a couple of years of exploitation, when the land had been cleared and the former frontier was being connected to outside markets through transportation and commercialization, large agrarian investors claimed ownership over the settlers' land. Peasants lacked the necessary knowledge, resources, and connections to obtain a title – all things that investors had – and generally failed to defend their claim (LeGrand 2016, 63-sq). The result was large-scale accumulation, with concessions of more than 1000 hectares representing more than three-fourths of the public lands titled between 1827 and 1931 (LeGrand 2016, 81).

Land accumulators had diverse social backgrounds. While some of them were members of the old landed classes, many were traders and financiers with a much more urban pedigree. Their common feature was their ability to use political connections to secure their claims to property. When they lacked the necessary networks, they could pay for them, as various law firms in Bogota offered their intermediation services to these 'frontier entrepreneurs'.[1] In addition to this land rush, their efforts were determined by attempts to secure access to labour, which required independent settlers to be transformed into service tenants or labourers (LeGrand 1984). This ruthless logic was summarized by early 20th-century peasant leader Erasmo Valencia when he said that landlords 'grab huge surfaces of land, not with the intent of farming it and making it produce, but of preventing others from working it' (Londoño Botero 2011). Tenancy contracts reinforced this dependency, for instance, by forbidding tenants to plant coffee on their parcels (Bejarano 2011, 501). As such, the expansion of the agrarian frontier was shaped by the will to jointly control land and labour, in a clear instance of 'primitive accumulation' (Marx 1992, 873-sq). From the point of view of peasants, the market was no longer an opportunity, but an imperative (Meiksins Wood 2002, 101).

Social unrest would ultimately result from such brutal forms of accumulation. However, until the early 20th century, these struggles remained scattered and localized (Palacios 2011; LeGrand 2016). This would change in the 1920s and 1930s, when land conflicts became much more politicized. They gained national visibility. They asserted a powerful claim: as land was accumulated far in excess of the landlords' production capacity, mobilized peasants argued that the ownership of public lands should be solely determined by labour. The land should belong to the ones who were labouring it.

To whom does the land belong?

These mobilizations were propelled by a favourable context. Through the 1920s, Conservative governments became more directly involved in the regulation of social conflicts. Drawing on International Labour Office (ILO) guidelines, new labour legislation was introduced (Palacios 2011, 124). These changes carried the supposed promise of a more protective state attitude towards service tenants, who often suffered under the yoke of the gruelling obligations imposed by their landlords. This was perceived by peasants as a window of opportunity. Here, the existence of brokers between the political sphere and peasant societies was critical to shaping the very idea of a possible common struggle. Some of these brokers were peasant leaders: people like Erasmo Valencia, who created the first 'agrarian' newspaper, *Claridad*, in 1928 and was the founder and leader of the National Agrarian Party (Londoño Botero 2011). Along with Valencia, hundreds of anonymous men and women contributed to a process of politicization of the peasantry, which gained visibility with the formation of collective movements such as 'agrarian leagues' (Sánchez 1985). Other brokers did not originate in rural societies. The urban leaders of leftist movements such as the Socialist Revolutionary Party and the

Communist Party became convinced that land conflicts were a momentous place for social struggle (Sánchez 1976). The most prominent of these brokers was Jorge Eliecer Gaitán, a lawyer, Congressman, and charismatic leader of the Liberal Party's leftist wing. He supported the struggles of peasants in Cundinamarca and Tolima, defending them in court and campaigning across rural municipalities (Londoño Botero 2011).

An additional factor transformed the political elites' approach regarding agrarian issues. The 1920s were a decade of accelerated industrial development in Colombia (König 1997). One of the consequences of rural migration and urbanization was the increase in food prices, which became endemic during that decade (López 1975). While capitalist agrarian entrepreneurs would invest in export crops, the country's food supply massively depended on peasants. This posed the question of land distribution. As food production needed to be stimulated, there was growing bipartisan agreement on the urgency of promoting peasant agriculture in order to feed the cities (Fajardo 1986, 39).

Things became even more complex in 1926, when the Supreme Court issued a ruling that intended to set clear principles for the recognition of property rights. In the judges' view, as Colombia was a settler nation, all landowners should be able to provide the proof that their estate had become private property in the past, through an adjudication decision. Otherwise, their estate was to be considered public land and could be allocated by the state to someone else (LeGrand 2016, 152–3). For peasants, this was a revolutionary decision. Encouraged by political activists and peasant leaders, tenants massively stopped paying their rents, claiming to be independent settlers as their landlords had no legal ownership rights (Londoño Botero 2011). This was a tremendous challenge to most landowners' claims and a threat to the very foundations of the social order. Interestingly, there was no clear-cut opposition among governing elites to the Court's decision. Instead of trying to reverse it, Congress voted in 1927 in favour of a mandatory revision of titles for all estates above 2500 hectares (LeGrand 2016, 153). Without any doubt, the proprietarian social order had never been – and would never again be – so fragile.

The stage was set for profound change when the 1929 global crisis hit Colombia. The areas most connected to the international economy through the production of export crops were particularly vulnerable (LeGrand 2016, 170-sq). Not only were rural employment and salaries drastically reduced, but many unemployed rural migrants returned to the countryside. Some of them had been in contact with labour unions and leftist movements, and became vectors of politicization (Fajardo 1986). While the Conservative regime attempted to respond to the challenge, these multiple crises propelled the candidacy of Liberal politician Enrique Olaya Herrera, who put an end to more than four decades of Conservative rule in the 1930 presidential race (Henderson 2001; Guerrero Barón 2014).

Once in power, the Liberal administration sought to scale down land conflicts. Pursuing a similar strategy to that of his Conservative predecessor, Olaya opted for the *parcelaciones* (parcelization) policy. This consisted of purchasing a large estate and dividing it among peasants. Parcelization had the support of landowners, as it provided the possibility of selling-off disputed estates and reinvesting the money into other activities (LeGrand 2016, 215). The first parcelization transactions were directly financed by the state, but the cost was prohibitive. From 1933, a new parcelization programme was initiated, with the Agrarian Bank purchasing land from its nominal owners and reselling it to the occupants (Palacios 2011, 113). Agrarian leaders harshly criticized this commercial shift. For them, this was land that peasants had previously cleared and made productive only to see it grabbed by the landlords. Not only had they been dispossessed, but they were forced to purchase the fruit of their own labour (Londoño Botero 2011; LeGrand 2016, 216). In spite of these criticisms, parcelization had a positive effect on land conflicts. Between 1935 and 1943, 217 estates, accounting for more than 90,000 hectares, were purchased and redistributed to 9947 beneficiaries (Londoño Botero 2011).

The government's next priority was to solve the legal conundrum posed by the differentiation of public and private lands and by the claims of both settlers and landlords over the same territories. After an unsuccessful attempt to pass an agrarian reform in 1933, a new project was retabled in July 1935 by President Alfonso López Pumarejo. While López Pumarejo was elected on a radical platform, his land bill was a mild compromise between the interests of peasants and landlords, and was to the clear advantage of the latter (LeGrand 2016, 231-sq). The bill, which became law in 1936, pursued the primary objective of scaling down the land conflicts that had raged for years or even decades. Differential treatment was given to settlers having occupied large private estates before and after 1935. For pre-existing conflicts, conciliation and land allocation procedures were possible. Newer occupations, on the contrary, were to be treated as squatting. The bill also sought to definitively close the door to incessant doubts about the differentiation between public lands and private property. An estate that had been a private possession for 30 years was considered private property, without consideration for the act originally granting ownership of it, its size, or the identity of the claimant. This provision consolidated support for landlords who had accumulated large tracts of land in the frontier areas. While the law endorsed the view that property rights could be lost if the land was not given a productive use, the provision remained toothless. Lastly, contrary to the failed 1933 bill, no limitation of the surface of rural estate was introduced.

The moderate inclination of the 1936 law reflects the balance of power at the time of the vote; leftist mobilization was declining as the parcelization policy had responded to many pressing demands (Fajardo 1986; Palacios 2011; LeGrand 2016). Moreover, while landlords had maintained a rather diffuse political stance in the 1920s, they became more organized and politically articulate during the 1930s. The influence strategy during the bill's debate was organized by the Farmer's Association (SAC), the Coffee-Growers' Federation, and the National Employer's Association (APEN) (Londoño Botero 2009). This reactionary bloc's concerns were voiced in Congress not by the Conservative Party, but by an influential rightist wing within the Liberal Party, which stigmatized any attempts at redistribution as having a 'communist' orientation (Henderson 2001; Londoño Botero 2009).

Over the following years, land policies were not innocuous; recent studies estimate that 14,043 plots, with an average surface of 54 hectares each, were distributed between 1930 and 1946 (Sánchez *et al.* 2020, 260). However, these policies were focused on formalizing the possession of public lands on the agrarian frontier and did not represent a threat of any kind for powerful landed interests. Overall, the 1930s land conflicts did not lead to the emergence of an organized peasant movement, as even the most structured agrarian organizations were never able to formulate sweeping claims rising above their specific struggles. In spite of the belief that many of the peasant leaders had in the transformative power of the law, the 1936 Lands Act greatly contributed to the consolidation of a proprietarian social order. Never again would the sanctity of property be called into question as it had been during the 1920s and 1930s.

The failed promises of peace

These years' multiple social tensions did not result in revolution, as many feared, but in nationwide bloodshed perpetrated in the name of partisan (Liberal versus Conservative) identities. Violence escalated after 1946, as Colombia was progressively engulfed by a civil war that led to 200,000 deaths and more than 2 million internal refugees (Meertens 2000, 228; Villamizar 2018, 178), a period that would later be known as *La Violencia* – the Violence. The war opposed the Conservative Party, which returned to power in 1946 and had the support of the police, the army, and multiple paramilitary militias, against a myriad of Liberal and Communist guerrilla groups. However, as often with civil wars (Kalyvas 2006), this partisan vocabulary concealed countless underlying economic and social conflicts at the subnational level. Agrarian conflicts were critical to cementing loyalties and radicalizing the opposition. In Cundinamarca or Tolima, where agrarian leagues and peasant unions had been influential since the 1920s, they provided the social structure for Liberal and Communist guerrillas (Medina 2007; Londoño Botero 2011). In some cases, partisan rivalry was a cover for a landlord's reaction against settler communities. But more frequently, violence spurred small-scale land grabbing, with peasants being dispossessed by people of similar social condition but opposing partisan affiliation (Ortiz Sarmiento 1985; Betancourt and García 1990; Roldan 2002).

In 1957, the leaderships of the Conservative and Liberal parties came to a transitional agreement to put a halt to violent conflict. To ward off partisan violence, the two parties would share assemblies and bureaucratic positions in equal proportions, and would alternate at the Presidency for a period of 16 years (Caballero Argáez *et al.* 2012). While the National Front – as this agreement was known – would later be harshly criticized as antidemocratic and elitist, its peacemaking ambitions would stir great hopes in its early years. The 1957 agreement and 1958 elections – the first democratic and reasonably warless race since 1946 – were seen at the time as the dawn of a 'Second Republic', and were followed by multiple efforts in favour of peace and reconciliation (Karl 2017).

Agrarian issues were paramount to the developmentalist peacemaking programme promoted by the first National Front president, Liberal statesman Alberto Lleras Camargo. Most of the victims of the 1950s violence were rural dwellers, many of whom had abandoned their plots and villages to seek refuge in the cities. The government's priority was to facilitate their return to the countryside and the recovery of their livelihoods. This was first done through an ambitious 'rehabilitation' programme, which included providing loans through the Agrarian Bank and building new infrastructure in rural areas (Karl 2017, 2018). However, this peacebuilding agenda fell victim to attacks from two directions. At the local scale, while new state involvement across rural territories was praised by rural communities, the benefits of the programme and the return of refugees spurred hatred and rivalries (Karl 2017, ch. 4). At a national scale, developmentalism became enmeshed in the complexities of coalition politics, as Rehabilitation became the preferred target of hard-line Conservatives. Lleras Camargo's successor, the regressive and arguably incompetent Conservative administration of Guillermo León Valencia, would do away with the programme's most ambitious aspects, while using its budget for patronage purposes (Karl 2018).

Post-conflict political economy was not only a matter of material rivalries, but also of frustrated expectations. One of the war's legacies was the existence of a close-knit network of communist communities, who settled in the mountains and deep valleys of the Southern Andes to escape from the military and aggression from rival Liberal guerrillas (Pizarro Leongómez 1991; Aguilera Peña 2014). After 1958, most of these groups welcomed the National Front's peace offers and agreed to lay down their weapons and collaborate with the state (Karl 2017). However, they became the target of violent anti-communist propaganda promoted both by the Conservative Party and by reactionary factions among the Liberals. These communities would later become the backbone of the FARC guerrilla group. However, turning against a teleological view of history, recent historical research has shown that, in the late 1950s and early 1960s, these men and women were not actually pursuing revolutionary aims. In reality, their demands were far more prosaic. They were seeking inclusion in the state and the market, pursuing a classic rural ideal of prosperity and solidarity, which the state would ultimately deny (Londoño Botero 2016, 98; Karl 2017, 88). Even in 1964, when southern Tolima Communists were remobilizing against what they perceived as a state-led aggression, their claims, gathered together in the guerrilla group's 'Agrarian programme', addressed both down-to-earth demands for 'an extensive credit system with payment facilities' and revolutionary perspectives for the future, such as 'the destruction of the old *latifundia* structure of Colombia' (Karl 2017, 198).

As such, the relationship between 1950s violence and the inception of revolutionary guerrilla groups is far more complicated than some observers contend (Richani 2002; Leech 2011). The 1960s were a decade of political agitation, which would eventually unfold into the creation of revolutionary leftist guerrillas who were craving to follow the Cuban path. However, this turmoil did not primarily take place among the peasant communities of the Southern Andes, but on city streets and university campuses. The most adamant promoters of armed revolution were pro-Cuban and pro-Chinese groups that would eventually gather in the ELN (National Liberation Army) and EPL (Popular Liberation Army) guerrilla groups (Villamizar 2017). Most of these would-be revolutionaries had an urban background, and often had a complicated relationship with agrarian issues, as they sought to graft themselves onto pre-existing land struggles (Rendón 2014). Leftist insurrectionists, particularly among Maoist circles, were violently critical of the Communist guerrillas, whom they considered as lacking revolutionary foresight (Aguilera Peña 2014, 69). From an opposite stance, Communist Party cadres were much more circumspect about the perspectives of armed revolution in Colombia, but eventually went on to create the FARC, to a certain extent as a reaction to leftist agitation (Pizarro Leongómez 1991, 189).

In the aftermath of the *Violencia*, the ruling classes' attitude towards the agrarian issue was shaped by the perception of the peasantry as a dangerous class. From 1946, the leadership of both parties attempted to mobilize the rural poor in a bloody game of political chess. The violence of the 1950s did not unfold along class lines, as the rural poor were divided by partisan identities, a feature that prevented the transformation of social strife into open insurrection (Pécaut 1987). However, the experience of war led guerrilla fighters to articulate new demands for citizenship that would ultimately unfold within the context of the opposition between the FARC and the state (Karl 2017). In the midst of war, the elite had realized the fragility of their grip over their party's people. The way in which the agrarian issue would be handled in the following decades would be marked by the dual aspiration to both appease and control the peasantry.

Impossible reform

Many of the Communists' agrarian programme demands were far from utopian for the time. In the early 1960s, land redistribution was not a radical demand, but a widely endorsed policy across the developing world. In 1961, in Punta del Este (Uruguay), the US government, the FAO, and a number of Latin American governments called for the implementation of land redistribution in the hemisphere as a way of addressing rural poverty and modernizing agriculture (Petras and Laporte 1971; Janvry 1981). However, in Colombia, as in many other places, the opportunity for change in the 1960s was short-lived. While an agrarian reform law was enacted by Congress in 1961, only a decade later, a powerful backlash annihilated most of the progress that had been made in the meantime. There were attempts in the 1980s to resurrect the idea of redistribution, linking it with the restoration of social peace in a growingly war-ridden countryside. Yet, by the end of the decade, redistributive agrarian reform was no longer a credible policy objective.

Overall, the result of more than two decades of agrarian reform policy fell far short of the early expectations expressed in the 1960s. Between 1962 and 1987, 970,741 hectares of land, covering 4814 estates, changed hands through state intervention (purchase, expropriation, cession by its original owner, or restitution).

Of these, the expropriation component represented a meagre 62,458 hectares, corresponding to 262 estates. These numbers contrast with the formalization of public lands, which massively took place in marginal areas colonized by landless peasants and occupied by large landowners alike. Between 1962 and 1987, more than 10 million hectares of public land became privately owned.[2] This exemplifies the fact that in Colombia's land policy, the expansion of the agrarian frontier massively superseded any form of redistribution. While the distribution of public lands was one of the responses provided to peasant claims for land, formalization often nurtured land accumulation. As a matter of fact, estates over 200 hectares covered 47.3 per cent of the public lands formalized between 1961 and 2010, but corresponded to a mere 2.5 per cent of the beneficiaries (Faguet *et al.* 2020). These numbers conceal variegated effects across the territory. In a recent and very thorough study, Faguet, Sánchez, and Villaveces (2020) demonstrate the critical effect of pre-existing land structure on policy outcomes. In municipalities where land was massively concentrated by 1960, public land distribution only consolidated existing inequality, with landlords grabbing most of the formalized lands. In contrast, in those places where land was less concentrated prior to the 1960s agrarian reform, land formalization nurtured a more egalitarian ownership structure. The following pages explore the social and political mechanisms behind these numbers.

An agrarianist approach

When Alberto Lleras Camargo became the National Front's first president in 1958, agrarian reform was an urgent matter. The post-transition regime was facing new waves of agrarian discontent, due in part to the consequences of *La Violencia*. The return of refugees to the countryside had emphasized the various challenges faced by any agrarian policy – not only rural violence, but also lack of access to land, credit, and transportation (Karl 2017). Economic factors also contributed to the urgency of agrarian reform. From 1954, a downturn in international coffee prices had evaporated Colombia's foreign exchange reserves, requiring the implementation of import restrictions. This would transform the country's dependence on agricultural imports into a political problem. Colombia was a rural country, but it still imported basic foodstuffs like rice and essential raw materials like cotton (Hirschman 1963, 125). Similar to the 1920s, when the toll taken by high food prices on industrial development fostered criticism of large landownership, the 1950s balance of payments crisis called for resolute support for the peasant economy, as opposed, in many political discourses, to the unproductive accumulation of idle land in the hands of large landlords.

The international context, with the Cuban revolution and the US-sponsored Alliance for Progress, would also feed this sense of urgency. Fidel Castro's success, as well as his authoritarian turn, would embolden the old anti-communist discourse, particularly entrenched among the military and right-wing politicians. But anti-communism's effectiveness would rely on its capacity to resonate with the concrete political reality in Colombia (Karl 2016). At the 1960 congressional midterms, electoral success by the MRL (the Liberal Recovery Movement, and later, the Liberal Revolutionary Movement), a leftist splinter group from the Liberal Party that received 19.4 per cent of the vote in the House of Representatives, made it clear that there was leftist opposition to the National Front. While led by a Liberal patrician, Alfonso López Michelsen (son of López Pumarejo, the father of the 1936 land law), the MRL's success in many rural areas was possible thanks to the support of peasant leaders. That was the case of Juan de la Cruz Varela, a historic figure of the Sumapáz land conflicts, who ran as López Michelsen's substitute for the House (Londoño Botero 2011).

The combination of all these factors led to an open context for change, which resulted in the passage of an Agrarian Reform Act in December 1961 (Law No. 135). The bill was shepherded by President Lleras Camargo's cousin, Carlos Lleras Restrepo. It featured a number of ambitious provisions, and was praised in its time by Albert O. Hirschman, who visited Colombia in 1960 and 1962 as a part of his research in development economics (Hirschman 1963). The law created an Institute for Agrarian Reform (Incora) endowed with a significant budget of 100 million pesos per year, which corresponded to 4 per cent of the country's annual fiscal revenue. A differentiated expropriation procedure was created, with compensation depending on land use. To the dismay of large landlords who before would systematically under-report their estate's worth, compensation was based on self-declared values for tax assessment. Incora was also made responsible for the implementation of rural development programmes such as irrigation and infrastructure construction, as well as the formalization of public lands to family farmers (Hirschman 1963, 147–52).

In spite of the favourable context and the ambitious provisions, the law's full implementation would have to wait until 1966, when the Liberals' return to power brought Carlos Lleras Restrepo into the presidential palace. The new president devised a reformist alliance between the peasantry and the progressive elites of the Liberal Party. However, this strategy would require a strong peasant organization capable of being viewed as the representative of the Colombian peasantry as a whole. Such an organization did not exist. Lleras Restrepo rectified this situation, establishing the National Peasant Association (ANUC, *Asociación nacional de usuarios campesinos*) by presidential decree in May 1967.[3] At its beginnings, ANUC was a peculiar organization. Its offices were located inside the Ministry of Agriculture's headquarters and its leaders were on the official payroll. The Department of Peasant Organizations (DOC), a ministerial division, was in charge of enrolling the members, promoting the creation of local chapters and delivering training sessions (Zamosc 1986). At the same time, ANUC's creation unleashed a wide grassroots movement, to a certain extent built on a new cycle of agrarian activism that had begun in the early years of the National Front.

However, ANUC's limitations were also tangible from its inception. On the one hand, ANUC faced the daunting challenge of representing a very heterogeneous peasantry composed of dependent tenants, Indigenous communities from the central Andes, settlers from the Amazonian frontier, moderately well-off independent coffee growers, and poor peons. On the other hand, ANUC ran the risk of creating excessive expectations, as its means of action remained very limited. It was supposed to participate to Incora's governance, yet it had only two seats on the Institute's board, while 13 other delegates represented the political parties, the administration, and large agricultural entrepreneurs (Fajardo 1986; Zamosc 1986; Machado 2017).

Along with the creation of ANUC, Lleras Restrepo introduced a new, potentially cutting-edge provision: the transformation of tenants into full owners through the expropriation and division of rented estates. While several restrictions were introduced during the congressional debates, the new procedure was a massive attack on the property structures of the Colombian countryside. It was met with a brutal reaction, as many landlords, especially in the Caribbean hinterland, expelled their tenants to claim full ownership of their estates (Zamosc 1986, 67). The evictions propelled a peasant backlash coordinated by local ANUC committees. In 1969, the peasants inaugurated the use of the 'seizure' repertoire (considered as 'recoveries' or 'invasions' depending on the point of view). Forcing their way onto estates became a preferred method of eliciting a reaction from the state.

Uprooting land reform

In August 1970, the alternation rules of the National Front brought a new Conservative president, Misael Pastrana, to power. Through his first year in office, the danger posed by ANUC to the proprietarian social order had become palpable. The number of land seizures was on the rise. From a mere 47 seizures in 1970, the following year saw that number swell to 645 (Zamosc 1986, 75). The adoption of these disturbing tactics was linked both to local dynamics of radicalization – the expulsion of tenants by their landlords and the creation of private militias by recalcitrant landowners – and to a nationwide feeling of frustration with Incora (Machado and Meertens 2010). Although the Institute was more active than any time before (or after), expectations were so high that they were very difficult to fulfil.

Less visible to the general public, but equally concerning for the government, was the radicalization of ANUC's membership, especially in the Caribbean plains and the frontier regions. Through 1970 and 1971, the Maoist Marxist-Leninist Party and the Trotskyist militants of the Socialist Bloc gained influence within the movement. While contemporary observers held ANUC's leftist turn responsible for the wave of land occupation seizures, Zamosc (1986, 92) convincingly argues that ideological radicalization was partly the consequence of state repression of these actions. Under Pastrana, the government's attitude became more supportive of landlords and the president vocally considered land seizures as a subversion of public order. As observed by a number of collective action scholars (e.g., Fillieule and Combes 2011), repression often marginalizes the moderate wings of social movements, benefits the most radical activists, and favours confrontational ideologies.

The ideological and practical radicalization of ANUC drove a wedge in the alliance between the peasant movement and the reformist bourgeoisie, represented by the progressive wing of the Liberals that had supported the reform (Zamosc 1986). The Liberal Party was divided, and many former supporters of Lleras Restrepo's progressive stance called for moderation in the application of the reform (Gilhodès 1974). From early 1971, the government deployed a twofold effort to push back against ANUC's activism. On the one hand, it favoured the fragmentation of the movement, already spurred by its intrinsic heterogeneity and by the divisions generated by the spread of a leftist affiliation among the majority of the militancy. Government agents from the DOC, the ministerial division that had been critically supportive of ANUC's inception, actively backed the moderate wing, spearheaded by local federations in coffee-growing areas. The Communist Party supported this divisive strategy, happy to counteract its Maoist rivals' influence (Zamosc 1986, 103–4). On the other hand, harsh repression was used against the radical wing. The military was deployed to regions affected by large-scale land seizures; activists were abusively imprisoned and harassed. Groups of hitmen paid by landlords killed and tortured some leaders and their families. Violence was particularly pervasive in the departments of Sucre and Córdoba, where ANUC committees were stronger and leftist influence more widespread (Machado and Meertens 2010, 227–8).

While repressing ANUC and favouring discord within it, the government pushed for a radical change in the implementation of land policy. In mid-1971, the moderate Minister of Agriculture Emilio Valderrama, who had advocated for an accommodating attitude towards ANUC and the continuing pursuit of the reform, was replaced by hardliner Hernán Jaramillo Ocampo, himself a landlord. In October 1971, the former vice-president of the country's top agribusiness group was nominated as head of Incora. In January 1972, a bipartisan agreement known as the Chicoral Pact established a new orientation for agrarian policy. In exchange for an increase on real estate taxes, the government committed to introducing a number of safeguards that, in practical terms, rendered redistribution by expropriation nearly impossible to implement. The agreement was widely supported by the different currents of the Liberal and Conservative parties, including the leftist MRL. Among political elites, only a weakened Lleras Restrepo and former Minister Valderrama openly criticized the agreement (Villamil Chaux 2015, 125-sq). The promises were translated into legislation passed by Congress in March 1973 in spite of strong opposition from ANUC (Zamosc 1986).

The Chicoral Pact limited redistribution to the voluntary sale of their estates by large landowners, who would freely negotiate the transaction with Incora. Not only did this create a massive 'corruption trap', but it also closed the door to any significant change in the country's property structure (Fajardo 1986, 126; López Enciso and Kalmanovitz 2006, 338). In lieu of redistribution, the government offered peasants the perspective of state-supported colonization of the Amazonian frontier. Similar to other Andean governments (Gootenberg and Dávalos 2018), Pastrana would praise the Amazon as a 'land without men for men without land'. Over the following decades, Incora's main mission became to grant titles for land farther into the frontier (Torres 2018). However, not only did colonization fail, it laid the foundations for new cycles of violence in the Nation's marginalized areas (Jaramillo *et al.* 1986; Molano 2010; Ramírez 2011).

Colombian elites' reactionary turn was both justified and inspired by international development aid ideas. The US government's attitude had changed between the first (1961) and the second (1967) conferences of Punta del Este. In 1961, Washington had vocally supported the objective of a structural transformation in the rural societies of the hemisphere, only to later realize that such an attitude might undermine its imperial interests by ousting US allies from power. By 1967, the objective was no longer to redistribute land, but to 'promote a rise in the standard of living of farmers' by increasing food production (Petras and Laporte 1971, 382). This change reflected the growing influence of a productivist approach within USAID's bureaucracy. As the Rockefeller Foundation promoted 'Green Revolution' techniques, a set of chemical and technological fixes praised for having supported the modernization of agriculture in Mexico, India, and the Philippines, agricultural science became a credible substitute for agrarian reform. It entertained the illusion that development could bring about a solution to the intrinsic contradictions of agrarian capitalism (Perkins 1997; Patel 2012).

The international commitment to productivism was embodied in Bogota by Lauchlin Currie, a New Dealer economist turned World Bank adviser who had become a guru for the nascent field of Colombian development economics (Sandilands 1990). Currie had been consistently promoting a productivist agenda for Colombia since 1961, when he devised a development programme based on support for large agribusiness companies, the transformation of a share of the peasantry into an agrarian proletariat, and the urban migration of the rest of the rural poor (Brittain 2005). The reactionary turn of the 1970s established a favourable context for Currie's ideas. President Pastrana embraced Currie's plan and made it the foundation of his development programme. Currie's ideas would never be fully implemented, and the Green Revolution programmes would fall well behind their quantitative objectives (Berry 2017). However, these doctrines provided the ideological framework for the invisibilization of the agrarian issue. In the 1970s, the problem was no longer framed in terms of justice or inequality, but productivity. The distribution of credit, seeds, and chemical inputs was real, but it only benefitted medium- and large-sized producers already connected to the market (Janvry 1981; Fajardo 1986). For instance, Green Revolution techniques provided the foundation for the transformation of cattle estates in the Cesar Valley (Caribbean hinterlands) into Colombia's cotton belt. Not only were the gains captured by the region's large landlords, but their estates' expansion was carried out at the price of the dispossession of small producers, who would go on to provide the labour needed for the thriving cotton plantations (Britto 2020, 37-sq).

While the agrarian reform project and ANUC's voice were durably silenced in the 1970s, the 1980s and early 1990s saw the rise of new concerns in relation to the agrarian issue, as land became enmeshed with conflict and peacemaking. This relationship was not a new one for Colombian politics, a reality made clear by the 'Rehabilitation' policies of the late 1950s and early 1960s. However, there were new actors on stage. During the 1980s, renewed visibility of the agrarian issue was brought about by negotiations with guerrilla groups and the emergence of drug trafficking.

Land, guerrillas, and drugs

During the 1970s, successive governments adopted a repressive approach to social opposition. However, not only did they fail to curb it, but they ultimately fed the growth of the most radical groups. Throughout the late 1970s and early 1980s, old guerrilla groups like the FARC and newer groups such as the M-19 urban rebels progressively became a central security concern. The 1982 arrival of Belisario Betancur, a moderate Conservative and Christian Democrat, to the presidency set the stage for peace talks and a ceasefire. Along with his peace plan, President Betancur promoted a developmentalist approach to social unrest, which was pursued under the following administration led by Liberal Virgilio Barco (1986–1990). Moreover, Barco's administration was marked by the influence of a new generation of development economists, who were familiar with the idea of an intrinsic link between rural poverty, land inequality, and violence (Pardo Rueda 1996). For the first time since Lleras Camargo's reconstruction policies, the state's incapacity – or unwillingness – to address this challenge was conceptually associated with the existence of a violent threat.

The flagship policy of this developmentalist turn was the National Rehabilitation Plan (PNR), a local development scheme focused on the most vulnerable and violent regions of the country. The PNR was deployed in 177 municipalities characterized by widespread poverty, poor infrastructure, and high levels of armed violence (Tirado 1990). While the PNR's main emphasis was infrastructure, the land distribution policy was significantly strengthened by the conceptual nexus made between rural poverty and war. According to official figures, Incora distributed close to a 100,000 hectares to landless peasants between 1983 and 1987. An additional 185,760 were distributed to demobilized guerrilla fighters and their families (Londoño Botero 2016, 160).

With the aim of bolstering agrarian reform policy, the Barco administration endeavoured to reform existing legislation, eliminating the multiple obstacles to redistribution established by the 1970 anti-reform policies. In March 1988, in spite of an aggressive reaction from Conservatives in Congress, the government managed to gain approval for Law No. 30. The text facilitated the expropriation of unexploited land, thus returning to the spirit of the original vision of the 1961 law. However, the transformative ambition of the law remained ambiguous, as it established higher compensation thresholds for expropriated landlords. In practice, expropriation remained extremely rare. Between 1988 and 1994, only nine estates, accounting for a mere total surface of 1256 hectares were subject to a successful expropriation procedure (Londoño Botero 2016, 163).

At this point, the whole debate on expropriation intersected with the growing public concern over massive land purchases by drug smugglers. During the second half of the 1980s, the government hardened its position towards drug traffickers, which led to an open war between the state and a segment of the criminal milieu. Within this context, the expropriation of drug lords' assets was conceived as one of the potential ways of halting their expansion and countering their political and social influence. Between 1989 and 1990, the Barco government issued a series of decrees providing the legal footing for the expropriation of real estate purchased with drug trafficking proceeds. In the same vein, it created a specialized agency, the National Narcotics Directorate, tasked with managing the confiscated properties.

Notably, in line with the law-enforcement-related justification for the measures ('hitting drug barons where it hurts – their finances'), the expropriation procedure was given a redistributive reasoning. The confiscated rural properties were meant to be transferred to Incora, which would be in charge of parcelling and allocating the land to peasant families. This responded to an underlying criticism of the drug economy. According to anthropologist Juana Dávila (2018, 144), from at least 1989, top Incora executives argued that the purchase of rural land by drug traffickers had the notable perverse effect of aggravating the already extremely unequal ownership structure in many rural territories. Initial estimates were staggering, ranging between 1 and 4 million hectares purchased (Dávila 2018, 145–9). In 1990, economists estimated that somewhere between 8 and 23 per cent of total drug profits repatriated in Colombia were being used to purchase land (Sarmiento and Moreno 1990, cited in Steiner 1999). In 1997, sociologist and jurist Alejandro Reyes produced the first methodologically grounded work in the field, estimating that over one-third of the country's municipalities presented serious indicators of land investment by drug entrepreneurs (Reyes 1997).

The way in which the debate on land owned by drug lords was framed is evocative of how the very idea of redistribution had changed over the 1970s and 1980s. While during Carlos Lleras' presidency, there was a rather broad consensus as to the necessity of enacting redistributive reform and the state's authority to implement it, by the 1980s land redistribution was seen as a radical idea that could only be justified when it affected the proceeds of ill-gotten gains. Admittedly, grabbing the narcos' lands was a way of pursuing expropriation without hurting the traditional dominant rural classes (Gutiérrez and Marín 2018). This is indicative of both the security policies of the time, and the manner in which the expropriation of the rural rich, in the sole name of the common good, had become unthinkable.

Identity politics and class politics

There was one exception to this general backlash against redistribution. While the traditional idea of redistributive land reform had been eradicated by years of regressive agrarian policy, an apparently contradicting vision made its way into the circles of power: the idea of territorial rights, i.e., the fact that a community could have exclusive rights over a certain territory, justified in the name of its social cohesion, cultural bonds, and historical occupation. Nevertheless, not just any rural inhabitants could aspire to these kinds of rights. A large range of actors – government officials,

NGO professionals, development assistance personnel, etc. – saw territorial rights as intrinsically associated with ethnicity. As elsewhere in Latin America, the last decade of the century spurred new approaches to ethnic others, and first and foremost Indigenous peoples.

This section will briefly sketch an analysis that would merit far more comprehensive treatment: the establishment of a right to land for ethnic subjects entailed the dismissal of the same rights for non-ethnic subjects. In the specific conditions of the last three decades in Colombia, ethnic politics have succeeded in dismissing class politics. First, I will show that the progressive inclusion of social and economic rights in the 1991 Constitution did not threaten the proprietarian regime in the countryside, but instead defined different rights for ethnic and non-ethnic subjects. Second, I will reflect on the ethnicization of territories that resulted both from social struggles and from politically framed opportunities. Last, I will show how peasants attempted to make a similar claim of a right to territory, but with a very different outcome.

The promises of the 1991 Constitution

In 1989 and 1990, the convergence of a broad student movement, civil unrest in response to the escalation of drug-related violence, and peace talks with several rebel groups – chief among them the urban guerrillas of the M-19 – resulted in a National Constitutional Assembly (ANC) tasked with the mission of drafting a new constitution, the first since 1886. The 1991 Constitution would become, for a whole generation of activists, progressive politicians, and modernist intellectuals, the foundation for a new social pact that was expected to establish the basis for a more peaceful and civil social order (Lemaitre Ripoll 2009). While such high expectations were impossible to fulfil, the new bill of rights introduced radical changes into the Colombian legal order, and has been praised as the most ambitious constitutional text of the hemisphere in terms of human rights, multiculturalism, and guarantees for the rule of law (Cepeda-Espinosa 2004; Noguera-Fernández and Criado de Diego 2011). This is largely due to the characteristics of the Assembly. The ANC was incredibly diverse in terms of political and social background and ethnicity, albeit not in terms of gender. It was dominated by a reformist alliance of the Liberal Party and the leftist Alianza Democrática M-19, the party representing the demobilized M-19 guerrilla fighters, and included representatives from Indigenous peoples and the leftist Unión Patriótica, a party initially associated with FARC combatants.

However, the comprehensive character of the 1991 Constitution's provisions on multiculturalism, individual freedom, and human rights contrasts with its recognition of the limited role of the state in economic matters. As very thoroughly analysed by legal sociologist César Rodríguez-Garavito (2011), the Constitution is the result of a compromise between two reformist projects, both of which were well represented inside the ANC and among the close advisers of the president. The first was a neoliberal project promoted by economists in top administrative roles who pushed for an open market agenda (independence of the central bank, privatization of state-owned corporations, liberalization of capital controls, etc.). The second was a 'neoconstitutional' ideal promoting civil, social, and economic rights, in addition to a judicial review mechanism (a strong Constitutional Court) intended to enforce them.

As such, while the Constitution embraces the responsibility of the state in promoting equality and reducing poverty, it effectively limits its own capacity to redistribute property. With the notable exception of the territorial rights of ethnically defined social groups, the Constitution's drafters made no attempts to go beyond the classic principle of the inviolability of ownership (Peña Huertas *et al.* 2014). In matters of property, and more specifically landed property, it recognized the century-old process of consolidation of a proprietarian social order and did nothing to balance it. This amounted to an invisibilization of peasants and peasant struggles concealed behind the notion of economic liberty and the free market. As analysed by Peña Huertas, Parada, and Zuleta (2014, 130), 'the Constitution seems to have forgotten the peasants. Throughout its 380 articles, they are only mentioned in article 64, where they are equated to "agricultural workers", which in reality does not correspond to being a peasant.'

This invisibilization contrasts with the recognition of Indigenous peoples' rights to the collective ownership of their 'ancestral lands', providing the legal base for the consolidation and extension of Indigenous reserves (*resguardos*). A 1993 law, fulfilling a constitutional mandate, extended this recognition to 'Afro-Colombians', defined as another ethnic group. Communities recognized by the state as authentic Indians and Blacks were granted a right of previous and informed consent on policy matters affecting them (Chaves 2011). Of course, these legal provisions often remained an official illusion, seldom protecting Indigenous and Black communities from the brutality of war. Indians and Blacks remained massively represented among the poorest of the rural poor. However, multiculturalism established a pattern of ethnically shaped (in)visibility that would result in differential treatment of claims and suffering.

Multiculturalism and ethnic territories

Of course, Colombia was not alone on the path of multiculturalism. During the final two decades of the century, most Latin American countries would implement multicultural reforms to their institutional design. These reforms were oriented towards the recognition of the cultural and territorial rights of Indigenous peoples, but also – in some cases – Afro-Latin communities. Van Cott (2000) and Yashar (2005) consider that these reforms pursued the ideal of democratization within post-conflict or post-authoritarian contexts. While deep changes were introduced in Latin American polities as the rights of hitherto marginalized and discriminated Indigenous populations were recognized, Hale (2002) argues that this multilateral turn was made compatible with a parallel and even more sweeping neoliberal turn. Multiculturalism has not been solely a vector of inclusion; it has also served as a disciplinary device, as it provided the background for selective inclusion, enabling the recognition of Indigenous peoples as partners in policy action while weakening the challenges represented by their most contentious representatives.

Of course, this new set of rights was not just handed over by the state in the late 1980s and early 1990s; it was the result of a long history of social struggles, which in many cases were intimately linked to the problem of land. In Colombia, the mobilization of the Indigenous communities of the Cauca mountains dates back to the 1910s struggles of Quintín Lame, the first contemporary leader of the Indigenous cause (Hristov 2005). A more recent and direct precedent can be found in the Indigenous struggles of the 1970s, which led to the creation of the Regional Indigenous Council of Cauca (CRIC) in 1971 (Troyan 2015). In both cases, land was the core issue of the mobilization.

State policy towards Indigenous peoples over the 19th and 20th centuries had consistently adopted an assimilationist approach. One of the consequences of this stance was the progressive dissolution of colonial-era reserves and their transformation into private property transferred to members of the community. Yet, these smallholders were easily dispossessed by neighbouring landlords and transformed into tenants or peons. The fight for the recognition and the extension of the reserves was the central struggle of the CRIC throughout the 1970s and 1980s (Findji 1992). Initially associated with ANUC, the CRIC would use similar strategies of land seizures and civil resistance. However, the CRIC was much more successful through its use of a legalistic repertoire. While this did not protect the movement from violent repression from state and local landlords, it faced the government with its own contradictions, thus becoming a powerful lever for change (Troyan 2015). Moreover, at least until the 1990s, Indigenous organizations insisted on the restoration of their reserves along the legal boundaries set by colonial titles (Findji 1992, 121). In a country in which colonial occupation acts as the philosophical foundation for all property, this claim was difficult to controvert.

In the 1980s, 21.8 million hectares of land were recognized as belonging to Indigenous reservations (Houghton 2008, 85). This, of course, was a tremendous victory for the Indigenous movement. It should, however, be nuanced, considering that much of this surface is located in marginalized regions of the country such as the Amazon and Pacific lowlands, where the recognition of collective rights seldom went against competing commercial interests. The recognition and expansion of reserves in areas of agrarian capitalism, as in the case of the Zenú people of the Caribbean plains, has been much more contentious (RECAR 2007). In any case, achieving this result required a sustained effort by the Indigenous movement, as well as a favourable political environment made possible by international interest in Indigenous people's rights (Van Cott 2000), increasingly associated with environmental issues, as well as by the work of government officials defending an 'indigenist' agenda within the state (Houghton 2008, 87; Troyan 2015).

The mobilization of Black people in favour of the recognition of comparable collective rights is much more recent. From the 1970s, the Black movement began to coalesce around claims linked to citizenship and discrimination. However, this was an urban movement that was primarily led by students and intellectuals. Its rural and territorial turn was largely triggered by Indigenous peoples' claims over the Pacific lowlands, where rural Black communities form the majority of the population (Wade 1993; Pardo 2002). The work of the Constitutional Assembly strengthened the politicization of Black communities around the issue of land. While existing Black movements failed to gain a seat in the ANC, they maintained sustained pressure on Assembly members through public protests relayed by Indigenous Assembly members, who contributed to framing the Afro-Colombian issue as a part of a larger multicultural agenda (Wade 1994; Van Cott 1996). However, the new Constitution did not settle the issue, and it was not until 1993 that a law recognizing the territorial rights of Afro-Colombian communities was passed by Congress. Largely inspired by the legislation on Indigenous reserves, the 1993 act defined the territorial rights of 'Black communities' over community-held land. In doing so, it transformed Black Colombians into ethnic others, conflating race and ethnicity (Restrepo 2004; Ng'weno 2007).

The general shift towards the setting aside of Indigenous and Black lands consisted of an ethnicization of territories, naturalizing the political boundaries between ethnically defined groups (Bocarejo 2009, 308). The production of these 'multicultural landscapes' conflates ethnic typologies and topologies, as Diana Bocarejo (2014) elegantly puts it. Moreover, this process has been buttressed by further environmentalist considerations, as Indigenous peoples and Black communities are portrayed by NGOs and government officials as naturally suited to being wardens of nature, an expectation imbued with a culturalist and ethnocentric vision of people's relation to their environment (Bocarejo 2009; Cárdenas 2012; Ojeda 2012).

This ethnic territorialization has contributed to the political marginalization of the peasantry (Chamorro and Escobar 2006). Multicultural legislation represents peasants as lacking any specific identity or cultural bond with the territory. In conservation areas, they are frequently defined as a threat to the environment, as they are deemed to have a purely instrumental relation to land, as opposed to the cultural specificity of Indians and Blacks (Bocarejo 2009; Ojeda 2012). As such, the flip side of the protection of ethnic lands has been to consider that non-ethnic subjects are nothing more than market subjects. This has been nurtured by multiple strategies utilized against the backdrop of multicultural institutions (Chaves 2011): those of Colombian and foreign conservationist NGOs, who promote imagined representations of the link between ethnic identity and conservation, those of government officials, who tend to use ethnic divisions as a government tactic, and of course, those of Indians and Blacks, both leaders and ordinary people, who have learned to master the ethnic card in order to gain the upper hand in sometimes decades-old conflicts between themselves and their mestizo neighbours (Hoffmann 2003; Corredor 2014).[4]

The peasantry as a political subject?

Along with the territorial claims of Indigenous and Black movements, peasant organizations made similar requests over the 1980s. To a certain extent, these territorial aspirations were already present in the 1920s and 1930s parcelizations, which carried the promise of protecting smallholders from land accumulators. During the 1980s peace talks, peasants from the Caguán river basin – a FARC stronghold – requested the support of the state to implement land use planning measures in order to stabilize settlements, protect vulnerable natural zones, and support peasant agriculture (Ordóñez Gómez 2012, 18). Similar requests were made by the peasant communities of the Macarena mountains – another marginalized FARC stronghold – where local communities strived to tie together peacemaking, environmental conservation, and peasant agriculture (Fajardo 2014, 156). However, these initiatives were hindered by the end of the peace talks and the escalation in official and paramilitary violence.

The experience gained in these marginalized zones was brought up during the preparation of a new land statute, which the government started to discuss in 1992. The primary objective was to reform a very discredited Incora, but the legislation would also import the market-led agrarian reform paradigm, which had gained tremendous currency all over the Global South in previous years (Borras 2007). This paradigm, opposed to the old 'state-led model' was intended to correct market distortions and provide incentives to economic actors to stimulate the access of landless and land-poor peasants to an individual plot of land (Binswanger *et al.* 1993; Deininger and Binswanger 1999).

While policy discussions in Bogota were primarily focused on a technical reform of land legislation (DNP 1992), social mobilization in 1993 forced the government to consider alternative approaches to the land statute. In September, peasant and Indigenous organizations, with the support of workers' unions, launched an alternative project and called for a national strike if they were not listened to by the government. The Agriculture Minister was José Antonio Ocampo, a renowned economist and scholar with close ties to the social-democratic wing of the Liberal Party. He opened talks with the organizations and made two important promises: reshuffling the expropriation procedures for idle land and responding to peasants' territorial claims by creating peasant reserves (*Zonas de reserva campesina*, or ZRCs) (Mondragón 2002, 13). ZRCs are areas where land is reserved for smallholders, with legal provisions prohibiting the accumulation of surfaces over a certain limit and measures intended to formalize untitled plots and allocate land to landless or land-poor peasants. Unlike the Indian and Black communities' territories, land within the ZRC was individually owned. However, ZRCs were also tools for grassroots development, thus recognizing peasants' legitimate claims to the uses of the territory.

Ulterior debates on this legislative creation illustrate the obstacles preventing people defined as market subjects from gaining recognition over their right to land and territory. After the bill's approval, there was a struggle of influence between peasant organizations and large producers' lobbying groups. While the former sought to turn the ZRCs into a tool for redistribution, the latter contended that they were not well suited to commercial farming areas and should be limited to the agrarian frontier. The issue was ultimately settled with the signature of a decree and an administrative agreement in 1996, ostensibly paving the way for broad application of the law (Mondragón 2002, 20).

An important aspect of ZRCs was the fact that they acknowledged that peasants could be green subjects, just as had been the case for ethnically defined populations. As a matter of fact, one of the policy objectives pursued by ZRCs was the stabilization of the agrarian frontier in environmentally vulnerable zones, especially in the Amazon Andes. By protecting peasant populations from the market pressure that has traditionally pushed them further into the frontier, the ZRC would constitute a 'buffer zone' surrounding protected areas.

The first peasant reserve was established in 1998 in one of these buffer zones, the El Pato region, around Los Picachos Natural Park, a marginalized zone in the Amazon Andes, characterized by contentious colonization, guerrilla influence, and state repression. Preliminary studies were undertaken in two other zones, Calamar – also on the edge of the Amazonian rainforest – and Cabrera (a decades-old peasant mobilization spot in the central department of Cundinamarca). The programme was funded by the World Bank and implemented by Incoder with the support of the Ministry of the Environment (Fajardo 2014, 159–60). The mobilizations of 1993–1996, as well as the existence of a legal instrument supporting the peasants' territorial claims, stimulated the organization of peasant associations in other recently colonized areas such as the Cimitarra river association, created in 1996 (Méndez Blanco 2013). In the following years, these organizations came together to form the National Association of Peasant Reserve Zones (Anzorc), which became a platform for domestic and international advocacy and a vector for the professionalization of activism (Allain 2019).

However, after a few years of successful implementation, which saw the creation of six peasant reserves over a total territory spanning nearly 900,000 hectares (Ordóñez Gómez 2012, 28), the ZRCs fell victim to attacks from counterinsurgent hawks. In 2002, Álvaro Uribe, a Liberal dissident with a sulphurous reputation of proximity to paramilitary groups, was elected to the presidency on a hard-line platform against guerrilla rebels. The ZRC programme was rapidly shut down under the pretext that these zones could become havens for rebel activity (Ordóñez Gómez 2012; Méndez Blanco 2013). The remaining ZRCs survived under tremendous strain and pressure, but their extension into new territories was halted. The return of ZRCs would be included among the 2016 peace-agreement provisions.

The ethnicization of territorial politics and the difficulties encountered by the ZRCs are characteristic of a transformation in the representation of territories. The mobilization of Indigenous and Black communities against the backdrop of a favourable context for multiculturalism has resulted in the recognition of the fact that there are places where land is socially embedded in cultural norms and collective rights, and should consequently be protected from the market. However, legal provisions and political decisions have carefully circumscribed this representation of the territory to Indigenous reserves and Black community territories. I contend that the logic behind this bonding is intrinsically tied to the fact that ethnic territories are an exception to the general rule under which land is considered as an abstract and exchangeable asset, stripped of any ties to culture or citizenship (Polanyi 1944).

Conclusion

In this chapter, I have traced the historical formation of an inequality regime. The first section showed that a very unequal property structure is intrinsically linked to the inception of agrarian capitalism, the place of Colombia in the first globalization, and the expansion of the agrarian frontier. These conditions sparked a rush for land, as influential landlords and investors accumulated huge tracts of land that were often well beyond their own investment capacities. In doing so, they secured access to a labour reserve of landless and dispossessed peasants whose only option was to become service tenants or sharecroppers. This capitalist accumulation itself created the conditions for it to be challenged, as the first peasant movements in the early 20th century disputed the very foundations of property, challenging the ownership rights of the landed elite. However, mild concessions sufficed to push back the threat, as the peasantry remained a fragmented and heterogeneous social group. As it emerged from this early history of social struggles, the land inequality regime was characterized by a radical defence of property rights and a concealment of their dubious foundations.

All challenges to this proprietarian social order though the 20th century eventually failed. First, in the aftermath of the brutal civil war of the 1950s, land policies were framed as a critical component of peacemaking and reconciliation. However, this transformative stance failed to result in lasting change, and eventually went on to feed the frustrations of the rural poor, nurturing the claims of those who advocated for revolution. In the late 1960s, a new attempt to implement a land reform was frustrated by the reaction of the landed elites and the leadership of the dominant political parties, who realized the threat that an organized peasantry could represent for social order. In the 1970s, the very idea of land redistribution was done away with. Social advancement was reframed as a problem of production, not redistribution, as the technology of the Green Revolution offered the promise of overcoming the fundamental contradictions of agrarian capitalism. In the 1980s, the emergence of a variety of security threats to state authority, from guerrillas to drug lords, contributed to the return of agrarian reform in the policy debate. However, these discussions saliently omitted any form of redistributive state intervention, except in the case of the expropriated proceeds from ill-gotten gains. The association between expropriation, law enforcement, and redistribution only highlights how sacrosanct the principle of property had become by then. This chapter comes to a close on the topic of the last decade of the 20th century, as ethnicity became intrinsically associated with the principle of territorial rights. While the idea that a community's rights over a territory, by virtue of their cultural bonds and historical struggles, might seem like a radical challenge to the proprietarian social order, I have argued that its transformative potential has been restricted by its limitation to populations defined as 'ethnic others'. This has become a tremendously challenging obstacle to peasant mobilizations as they have strived to obtain similar rights.

The next two chapters will narrow this perspective in two ways. First, chronologically, they will deal with the contemporary armed conflict through the lens of one specific actor: paramilitary militias. Second, they will also narrow the view geographically, as they will be focused on a local case study – the northern part of the department of Magdalena, a place that embodies many of the characteristics of agrarian capitalism described in the present chapter. Together, they will allow me to unfold the core argument of this book on the relationship between violence and inequality regimes.

Notes

1 To paraphrase Eric Léonard (2017), who uses the term in the Mexican context. The social background and legal practices of land accumulators are described by LeGrand (2016).
2 All the figures are from Londoño (2016, 150–1), which is the most thorough exploration to date of the Incora archives.
3 The main source on ANUC remains the classic book by Zamosc (1986). This paragraph is based on Chapter 2 of his book.
4 I will not touch on the complexity of these interethnic land conflicts here, which can also oppose Indigenous and Black social groups. For a thorough view, see notably Chaves (2011).

References

Aguilera Peña, M. 2014. *Guerrilla y población civil: trayectoria de las FARC 1949–2013*. Bogotá: Centro Nacional de Memoria Histórica.

Allain, M. 2019. Les jeux d'échelles de l'action collective: militantisme local et solidarité internationale dans les campagnes de Colombie. *Critique internationale*, 82(1), 51–73.

Appelbaum, N.P. 2003. *Muddied Waters: Race, Region, and Local History in Colombia, 1846–1948*. Durham, NC: Duke University Press.

Bejarano, J.A. 2011. *Estudios de historia e historiografía. Historia agraria*. Bogotá: Universidad Nacional de Colombia.

Bergquist, C.W. 1986. *Coffee and Conflict in Colombia, 1886-1910*. Durham, NC: Duke University Press.

Berry, M.A. 2017. *Avance y fracaso en el agro colombiano, siglos XX y XXI*. Bogotá: Editorial Universidad del Rosario.

Betancourt, D. and M. García. 1990. *Matones y cuadrilleros: Origen y evolución de la violencia en el occidente colombiano, 1946-1965*. Bogotá: Tercer Mundo.

Binswanger, H., K. Deininger, and G. Feder. 1993. *Power, Distortions, Revolt, and Reform in Agricultural Land Relations*. The World Bank, No. WPS1164.

Bocarejo, D. 2009. Deceptive Utopias: Violence, Environmentalism, and the Regulation of Multiculturalism in Colombia. *Law & Policy*, 31(3), 307–29.

Bocarejo, D. 2014. Legal Typologies and Topologies: The Construction of Indigenous Alterity and Its Spatialization Within the Colombian Constitutional Court. *Law & Social Inquiry*, 39(2), 334–60.

Borras, S. 2007. *Pro-Poor Land Reform: A Critique*. Ottawa: University of Ottawa Press.

Brittain, J.J. 2005. A Theory of Accelerating Rural Violence: Lauchlin Currie's Role in Underdeveloping Colombia. *The Journal of Peasant Studies*, 32(2), 335–60.

Britto, L. 2020. *Marijuana Boom: The Rise and Fall of Colombia's First Drug Paradise*. Oakland, CA: University of California Press.

Bucheli, M. 2005. *Bananas and Business: The United Fruit Company in Colombia, 1899-2000*. New York: NYU Press.

Caballero Argáez, C., M. Pachón Buitrago, and E. Posada Carbó, eds. 2012. *Cincuenta años de regreso a la democracia: Nuevas miradas a la relevancia histórica del frente nacional*. Bogotá: Uniandes.

Cárdenas, R. 2012. Green Multiculturalism: Articulations of Ethnic and Environmental Politics in a Colombian 'Black Community'. *The Journal of Peasant Studies*, 39(2), 309–33.

Cepeda-Espinosa, M.J. 2004. Judicial Activism in a Violent Context: The Origin, Role, and Impact of the Colombian Constitutional Court. *Washington University Global Studies Law Review*, 3(4), 529–700.

Chamorro, M.C. and M.Z. Escobar. 2006. From Blanqueamiento to Reindigenización: Paradoxes of Mestizaje and Multiculturalism in Contemporary Colombia. *European Review of Latin American and Caribbean Studies*, (80), 5–23.

Chaves, M., ed. 2011. *La multiculturalidad estatalizada: indígenas, afrodescendientes y configuraciones de estado*. Bogotá: Instituto Colombiano de Antropología e Historia.

Christie, K.H. 1978. Antioqueño Colonization in Western Colombia: A Reappraisal. *The Hispanic American Historical Review*, 58(2), 260–83.

Corredor, J. 2014. Négociations identitaires et légitimités territoriales: les Chilapos dans le Bajo Atrato en Colombie. *Cahiers de l'Urmis*, (15). DOI: 10.4000/urmis.1261.

Dávila, J. 2018. A Land of Lawyers, Experts and "Men Without Land": The Politics of Land Restitution and the Techno-Legal Production of "Dispossessed People" in Colombia. PhD diss., Harvard: Harvard University.

Deininger, K. and H. Binswanger. 1999. The Evolution of the World Bank's Land Policy: Principles, Experience, and Future Challenges. *The World Bank Research Observer*, 14(2), 247–76.

DNP (Departamento Nacional de Planeación). 1992. *CONPES 2590. Nuevo impulso a la reforma agraria*. Bogotá: Consejo Nacional de Política Económica y Social.

Faguet, J.-P., F. Sánchez, and M.-J. Villaveces. 2020. The Perversion of Public Land Distribution by Landed Elites: Power, Inequality and Development in Colombia. *World Development*, 136, 105036.

Fajardo, D. 1986. *Haciendas, campesinos y políticas agrarias en Colombia, 1920-1980*. Bogotá: Centro de Investigaciones para el Desarrollo.

Fajardo, D. 2014. *Las guerras de la agricultura colombiana, 1980-2010*. Bogotá: ILSA.

Fillieule, O. and H. Combes. 2011. De la répression considérée dans ses rapports à l'activité protestataire. *Revue française de science politique*, 61(6), 1047–72.

Findji, M.T. 1992. From Resistance to Social Movement: The Indigenous Authorities Movement in Colombia. *In*: A. Escobar and S.E. Alvarez, eds. *The Making of Social Movements in Latin America: Identity, Strategy, and Democracy*. Boulder, CO: Westview Press, pp. 112–32.

Gilhodès, P. 1974. *Politique et violence La question agraire en Colombie*. Paris: Armand Colin.

Gootenberg, P. and L. Dávalos, eds. 2018. *The Origins of Cocaine: Colonization and Failed Development in the Amazon Andes*. London: Routledge.

Guerrero Barón, J. 2014. *El proceso político de las derechas en Colombia y los imaginarios sobre las guerras internacionales 1930–1945*. Bogotá: Universidad Pedagógica y Tecnológica de Colombia.

Gutiérrez, F. and M. Marín. 2018. Tierras en el posconflicto: ¿en el fondo cuál es el problema? *Análisis político*, 31(92), 18–38.

Hale, C.R. 2002. Does Multiculturalism Menace? Governance, Cultural Rights and the Politics of Identity in Guatemala. *Journal of Latin American Studies*, 34(3), 485–524.

Henderson, J.D. 2001. *Modernization in Colombia: The Laureano Gomez Years, 1889-1965*. Gainesville, FL: University of Florida Press.

Hirschman, A.O. 1963. *Journeys Toward Progress. Studies of Economic Policy-Making in Latin America*. Westport, CT: Greenwood Press.

Hoffmann, O. 2003. *Communautés noires du Pacifique colombien. Innovations et dynamiques ethniques*. Paris: Karthala-IRD.

Houghton, J. 2008. Legalización de los territorios indigenas de Colombia. *In*: J. Houghton, ed. *La tierra contra la muerte: conflictos territoriales de los pueblos indígenas en Colombia*. Bogotá: CECOIN, pp. 83–143.

Hristov, J. 2005. Indigenous Struggles for Land and Culture in Cauca, Colombia. *The Journal of Peasant Studies*, 32(1), 88–117.

Janvry, A. de. 1981. *The Agrarian Question and Reformism in Latin America*. Baltimore, MD: Johns Hopkins University Press.

Jaramillo, J.E., L. Mora, and F. Cubides. 1986. *Colonización, coca y guerrilla*. Bogotá: Alianza Editorial Colombiana.

Kalyvas, S. 2006. *The Logic of Violence in Civil War*. New York, NY: Cambridge University Press.

Karl, R.A. 2016. Reading the Cuban Revolution from Bogotá, 1957–62. *Cold War History*, 16(4), 337–58.

Karl, R.A. 2017. *Forgotten Peace: Reform, Violence, and the Making of Contemporary Colombia*. Oakland, CA: University of California Press.

Karl, R.A. 2018. From 'Showcase' to 'Failure': Democracy and the Colombian Developmental State in the 1960s. *In*: M.A. Centeno and A.E. Ferraro, eds. *State and Nation Making in Latin America and Spain: The Rise and Fall of the Developmental State*. Cambridge: Cambridge University Press, pp. 73–103.

König, H.-J. 1997. Los años veinte y treinta en Colombia: ¿Época de transición o cambios estructurales? *Ibero-amerikanisches Archiv*, 23(1/2), 121–55.

Leech, G. 2011. *The FARC: The Longest Insurgency*. London: Zed Books.

LeGrand, C. 1984. Labor Acquisition and Social Conflict on the Colombian Frontier, 1850-1936. *Journal of Latin American Studies*, 16(1), 27–49.

LeGrand, C. 2016. *Colonización y protesta campesina en Colombia: 1850-1950*. Bogotá: Tandem.

Lemaitre Ripoll, J. 2009. *El derecho como conjuro. Fetichismo legal, violencia y movimientos sociales*. Bogotá: Siglo del Hombre.

Léonard, É. 2017. Réforme agraire et reconfiguration du régime de gouvernementalité dans Les Tuxtlas, Mexique, 1920-1945. *Critique internationale*, 75(2), 53–69.

Londoño Botero, R. 2009. Concepciones y debates sobre la cuestión agraria (1920-1938). *In*: R. Sierra Mejía, ed. *República Liberal: sociedad y cultura*. Bogotá: Centro Editorial de la Facultad de Ciencias Humanas de la Universidad Nacional de Colombia, pp. 47–115.

Londoño Botero, R. 2011. *Juan de la Cruz Varela. Sociedad y política en la región de Sumapaz*. Bogotá: Universidad Nacional.

Londoño Botero, R., ed. 2016. *Tierras y conflictos rurales. Historia, políticas agrarias y protagonistas*. Bogotá: Centro Nacional de Memoria Histórica.

López, H. 1975. La inflación en Colombia en la época de los veintes. *Cuadernos Colombianos*, 2(5), 41–139.

López Enciso, E. and S. Kalmanovitz. 2006. *La agricultura colombiana en el siglo XX*. Bogotá: Fondo de Cultura Económica.

Machado, A. 2017. *El problema de la tierra: Conflicto y desarrollo en Colombia.* Bogotá: Penguin Random House.
Machado, A. and D. Meertens, eds. 2010. *La tierra en disputa. Memorias de despojo y resistencia en la Costa Caribe (1960-2010).* Bogotá: Centro Nacional de Memoria Histórica.
Marx, K. 1992. *Capital: A Critique of Political Economy. Volume 1.* London: Penguin Classics.
Medina, M. 2007. La resistencia campesina en el sur del Tolima. *In*: G. Sánchez and R. Peñaranda, eds. *Pasado y presente de la violencia en Colombia.* Bogotá: La Carreta, pp. 269–95.
Meertens, D. 2000. *Ensayos sobre tierra, violencia y género.* Bogotá: Universidad Nacional, Centro de Estudios Sociales.
Meiksins Wood, E. 2002. *The Origin of Capitalism: A Longer View.* London: Verso.
Méndez Blanco, Y. 2013. Derecho a la tierra y al territorio, justicia y zonas de reserva campesina el caso del Valle del Río Cimitarra. MA diss., Bogotá: Universidad Javeriana de Colombia.
Molano, A. 2010. *Selva adentro. Una historia oral de la colonización del Guaviare.* Bogotá: Punto de Lectura.
Mondragón, H. 2002. *Expresiones del campesinado.* Bogotá: ILSA.
Múnera, A. 2008. *El fracaso de la nación. Región, clase y raza en el Caribe colombiano.* Bogotá: Planeta.
Ng'weno, B. 2007. Can Ethnicity Replace Race? Afro-Colombians, Indigeneity and the Colombian Multicultural State. *The Journal of Latin American and Caribbean Anthropology*, 12(2), 414–40.
Noguera-Fernández, A. and M. Criado de Diego. 2011. La Constitución colombiana de 1991 como punto de inicio del nuevo constitucionalismo en América Latina. *Estudios Socio-Jurídicos*, 13(1), 15–49.
Ocampo, G.I. 2007. *La instauración de la ganadería en el Valle del Sinú: la hacienda Marta Magdalena, 1881-1956.* Medellín: Universidad de Antioquia.
Ocampo, J.A. 1996. *Historia económica de Colombia.* Bogotá: Tercer Mundo Editores.
Ojeda, D. 2012. Green Pretexts: Ecotourism, Neoliberal Conservation and Land Grabbing in Tayrona National Natural Park, Colombia. *The Journal of Peasant Studies*, 39(2), 357–75.
Ordóñez Gómez, F. 2012. *Zonas de reservas campesinas: elementos introductorios y de debate.* Bogotá: ILSA.
Ortiz Sarmiento, C.M. 1985. *Estado y subversión en Colombia: La violencia en el Quindío, años 50.* Bogotá: Fondo Editorial CEREC.
Palacios, M. 2002. *Coffee in Colombia, 1850-1970: An Economic, Social and Political History.* Cambridge: Cambridge University Press.
Palacios, M. 2011. *¿De quién es la tierra?. Propiedad, politización y protesta campesina en la década de 1930.* Bogotá: Fondo de Cultura Económica.
Pardo, M. 2002. Entre la autonomía y la institucionalización: Dilemas del movimiento negro colombiano. *Journal of Latin American Anthropology*, 7(2), 60–84.
Pardo Rueda, R. 1996. *De primera mano. Colombia 1986-1994: Entre conflictos y esperanzas.* Bogotá: CEREC/Norma.
Patel, R. 2012. The Long Green Revolution. *The Journal of Peasant Studies*, 40(1), 1–63.
Pécaut, D. 1987. *L'ordre et la violence: évolution socio-politique de la Colombie entre 1930 et 1953.* Paris: Editions de l'EHESS.
Peña Huertas, R., M.M. Parada, and S. Zuleta. 2014. La regulación agraria en Colombia o el eterno déjà vu hacia la concentración y el despojo: un análisis de las normas jurídicas colombianas sobre el agro (1991-2010). *Estudios Socio-Jurídicos*, 16(1), 121–64.
Perkins, J.H. 1997. *Geopolitics and the Green Revolution: Wheat, Genes, and the Cold War.* Oxford: Oxford University Press.

Petras, J. and R. Laporte. 1971. *Cultivating Revolution: the United States and Agrarian Reform in Latin America*. New York, NY: Random House.

Piketty, T. 2020. *Capital and Ideology*. Cambridge: The Belknap Press.

Pizarro Leongómez, E. 1991. *Las FARC (1949-1966): De la autodefensa a la combinación de todas las formas de lucha*. Bogotá: Tercer Mundo.

Polanyi, K. 1944. *The Great Transformation: The Political and Economic Origins of Our Time*. New York, NY: Farrar & Rinehart.

Posada Carbó, E. 1998. *El Caribe colombiano: una historia regional (1870-1950)*. Bogotá: Banco de la República.

Ramírez, M.C. 2011. *Between the Guerrillas and the State: The Cocalero Movement, Citizenship, and Identity in the Colombian Amazon*. Durham, NC: Duke University Press.

RECAR. 2007. Pueblo Zenú, recuperador de Sueños. Resguardo Indígena Zenú de San Andrés de Sotavento, Córdoba y Sucre. *Revista Semillas*, 32–33.

Rendón, J.C.V. 2014. "Los del campo", "los de la ciudad". Ideología organizacional, vanguardia revolucionaria campesina y aislamiento político del Ejército de Liberación Nacional, 1962-1973. *Análisis político*, 27(81), 49–63.

Restrepo, E. 2004. Ethnicization of Blackness in Colombia. *Cultural Studies*, 18(5), 698–753.

Reyes, A. 1997. Compra de tierras por narcotraficantes. *In*: F. Thoumi, ed. *Drogas ilícitas en Colombia. Su impacto económico, político y social*. Bogotá: Ariel Ciencia Política, pp. 279–346.

Richani, N. 2002. *Systems of Violence: The Political Economy of War and Peace in Colombia*. Albany, NY: SUNY Press.

Rodríguez-Garavito, C. 2011. Toward a Sociology of the Global Rule of Law Field: Neoliberalism, Neoconstitutionalism, and the Contest over Judicial Reform in Latin America. *In*: Y. Dezalay and B. Garth, eds. *Lawyers and the Rule of Law in an Era of Globalization*. London: Routledge, pp. 164–90.

Roldan, M. 2002. *Blood and Fire: La Violencia in Antioquia, Colombia, 1946-1953*. Durham, NC: Duke University Press.

Román, R., ed. 2014. *Economía del Caribe Colombiano y construcción de nación*. Bogotá: Universidad Nacional de Colombia.

Safford, F. 2010. El problema de los transportes en Colombia en el siglo XIX. *In*: A. Meisel and M.T. Ramírez, eds. *Economía colombiana del siglo XIX*. Bogotá: Fondo de Cultura Económica, pp. 523–73.

Sánchez, F., P. Torres, and M.-J. Villaveces. 2020. Tierra por votos. Adjudicación de baldíos durante la República Liberal, 1930-1946. *Revista de Economía Institucional*, 22(43), 249–75.

Sánchez, G. 1976. *Los bolcheviques del Líbano*. Bogotá: El Mohan Editores.

Sánchez, G. 1985. *Ensayos de historia social y política del siglo XX*. Bogotá: El Áncora.

Sandilands, R.J. 1990. *The Life and Political Economy of Lauchlin Currie: New Dealer, Presidential Advisor, and Development Economist*. Durham, NC: Duke University Press.

Sarmiento, L. and C. Moreno. 1990. Narcotráfico y sector agropecuario en Colombia. *Economía colombiana*, (226–227), 29–37.

Steiner, R. 1999. *Economic and Institutional Repercussions of the Drug Trade in Colombia*. Bogotá: CEDE, Universidad de los Andes.

Tirado, N. 1990. El Plan Nacional de Rehabilitación: un modelo institucional para la democracia participativa, la descentralización y la lucha contra la pobreza. *Coyuntura social*, (2), 121–43.

Torres, M.-C. 2018. The Making of a Coca Frontier: The Case of Ariari, Colombia. *In*: P. Gootenberg and L.M. Davalos, eds. *The Origins of Cocaine: Colonization and Failed Development in the Amazon Andes*. London: Routledge, pp. 133–59.

Troyan, B. 2015. *Cauca's Indigenous Movement in Southwestern Colombia: Land, Violence, and Ethnic Identity*. London: Lexington Books.
Van Ausdal, S. 2009. Pasture, Profit, and Power: An Environmental History of Cattle Ranching in Colombia, 1850–1950. *Geoforum*, 40(5), 707–19.
Van Cott, D.L. 1996. Unity through Diversity: Ethnic Politics and Democratic Deepening in Colombia. *Nationalism and Ethnic Politics*, 2(4), 523–49.
Van Cott, D.L. 2000. *The Friendly Liquidation of the Past: The Politics of Diversity in Latin America*. Pittsburgh, PA: University of Pittsburgh Press.
Villamil Chaux, C. 2015. *La reforma agraria del Frente Nacional: De la concentración parcelaria de Jamundí al Pacto de Chicoral*. Bogotá: Universidad Jorge Tadeo Lozano.
Villamizar, D. 2017. *Las guerrillas en Colombia: Una historia desde los orígenes hasta los confines*. Bogotá: Debate.
Villamizar, J.C. 2018. Elementos para periodizar la violencia en Colombia: dimensiones causales e interpretaciones historiográficas. *Ciencia Política*, 13(25), 173–92.
Wade, P. 1993. *Blackness and Race Mixture: The Dynamics of Racial Identity in Colombia*. Baltimore, MD: Johns Hopkins University Press.
Wade, P. 1994. Identités noires, identités indiennes en Colombie. *Cahiers des Amériques latines*, (17), 125–40.
Yashar, D.J. 2005. *Contesting Citizenship in Latin America: The Rise of Indigenous Movements and the Postliberal Challenge*. Cambridge: Cambridge University Press.
Zamosc, L. 1986. *The Agrarian Question and the Peasant Movement in Colombia: Struggles of the National Peasant Association, 1967-1981*. Cambridge: Cambridge University Press.

2

THE PATHS TO PARAMILITARY RULE

During the final two decades of the 20th century, Colombia was engulfed by an armed conflict of unprecedented intensity. Paramilitary militias played a critical role in the reproduction of the inequality regime described in the previous chapter. This was especially true in places where social mobilizations, economic transformations, and political changes represented acute threats to the dominant classes. But these groups were not simply a conservative force. They participated in the expansion of class exploitation in agrarian economies and beyond. In Lesley Gill's (2016) study on an industrial hotspot, they enabled the transformation of an organized proletariat into a precarious and manageable working force. In Teo Ballvé's (2020) study on an agrarian frontier, they were found to have disciplined the labour force and ensured the monopolization of land. In the case study I will analyse here, they forged ties between the control of electoral arenas, plantation labour, and land administration; they articulated physical violence, administrative power, and economic domination to expand and reproduce an exclusionary social order that was otherwise threatened by social and political mobilization. This violent enterprise was not alien to established political institutions, nor did it seek to undermine them. The paramilitary group's social networks encompassed local politics, business circles, and state agencies, and provided militia chiefs with the capacity to transform the fruits of violent coercion into legitimate assets (Grajales 2011; Ballvé 2012). As such, paramilitary groups cannot be studied as though they were criminal actors placed at the margins of institutions. The state was their field of action (Grajales 2017a).

This chapter will trace the formation of a paramilitary stronghold beginning in the early 1980s and over a period of 20 years. I will argue that paramilitary violence should be understood as woven into the structures of political and economic power at the regional and local level. These militias were neither the fruit of a rational counterinsurgency strategy governed by obscure interests, as some authors suggest (Avilés 2007; Hristov 2014; Zelik 2015), nor exogenous 'warlords' (Duncan 2005) simply driven by the geopolitics of drug trafficking. The trajectory of paramilitary militias is the result of a history of 'indirect rule' (Gutiérrez 2019), as these armed groups became key security providers for the local elites that have historically ruled at the subnational level.[1] As a result, paramilitary violence was not simply, and sometimes not primarily, the vehicle of a national anti-communist crusade, but more accurately the expression of local interests, as subnational ruling elites and owning classes were of course threatened by rebel activity, but also by leftist electoral mobilization and social activism. Seen from below, paramilitary militias were primarily an apparatus of repression, aimed at the maintenance of social order.

At the same time, as Francisco Gutiérrez (2019) argued, the interests of militias and local elites were not always perfectly aligned. The expansion of paramilitary groups near the start of the 2000s nurtured the appetite for power of armed militia leaders. At diverse degrees throughout the country, this unfolded into local strategies to accumulate political and economic power. More aggressive and proactive paramilitary leaders strived to set the conditions for local politicking and to seize control of local government institutions. This enabled them to access new sources of accumulation, as institutional influence was used to siphon public budgets and legalize land dispossession. While this expansive strategy harmed the interests of some powerful local clans, local patricians generally sought the militias' support in their competition with their peers, consolidating the paramilitary groups' position as central regulators of local power.

These arguments will be constructed on the basis of a local case study focusing on a typical militia stronghold. The department of Magdalena, located on the northern Caribbean coast of Colombia, gets its name from the eponymous river that delimits it to the west (see Map 2.1). With a surface of two-thirds that of Belgium, it has a population of roughly 1.2 million inhabitants, more than a third of whom live in the capital city, Santa Marta. The oldest city in Colombia and the third-largest commercial port in the Caribbean, Santa Marta is also one of the country's main tourist destinations. Another third of Magdalena's population lives in the neighbouring municipalities of Ciénaga, Aracataca, and Fundación. The north of the department, concentrating most of the population and producing the bulk of its revenues, is a strategic area. It is bounded by the sea, the Sierra Nevada de Santa Marta mountains, and the Ciénaga Grande wetlands, which extend westward to the Magdalena River. Land is extremely concentrated in Magdalena, with 10 per cent of the estates controlling 74.6 per cent of the farmland (DANE 2016).

The lowlands north of the department form what people there call the 'banana district' (*zona bananera*), an agribusiness area that has been connected to the globalized food market since the end of the 19th century. In 1899, the United Fruit Company (UFCO), a corporation that would spearhead US imperialism in the Caribbean, established one of its most productive banana-export businesses in this area. For over half a century, UFCO was the dominant actor in a transnational value chain integrating US investors, regional elites acting as local associates, as well as thousands of workers (Botero and Guzmán Barney 1977; Bucheli 2005; LeGrand 2012). At the end of the 1960s, the local banana economy plummeted, as

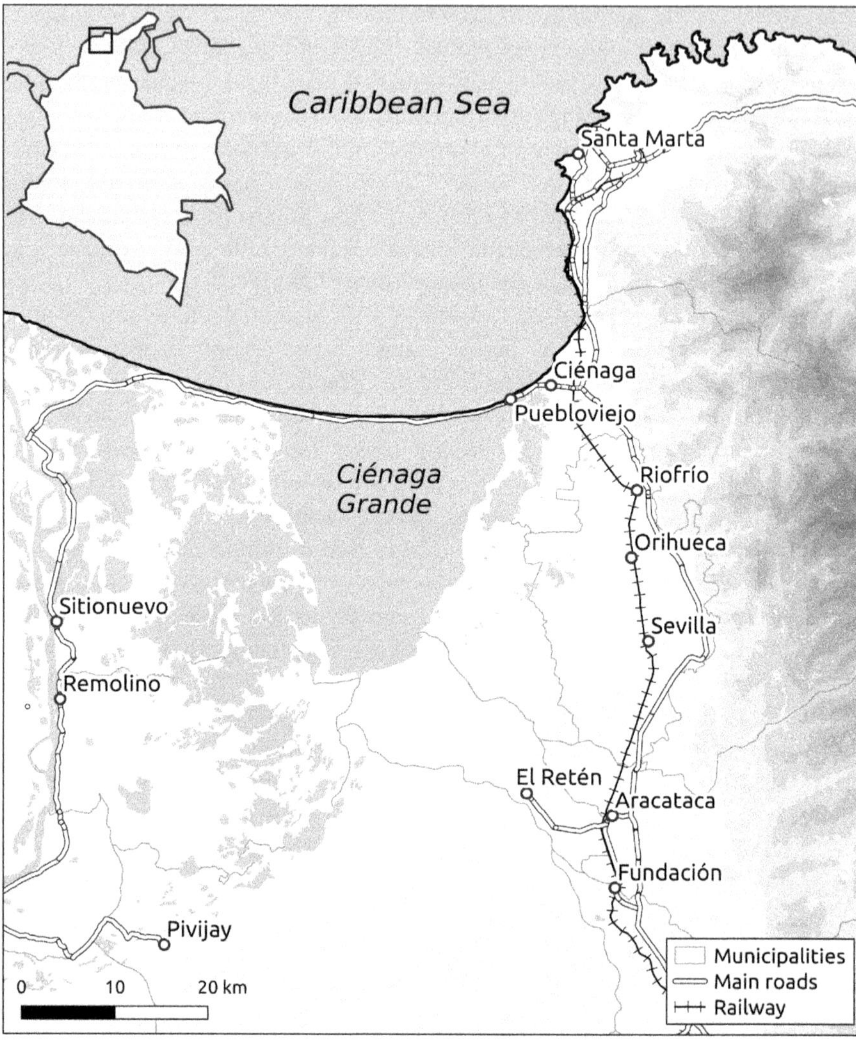

MAP 2.1 Northern Magdalena. Source: the author, created using data from OpenStreetMap (Open Database Licence – ODbL), OpenDem project (ODbL), and the Colombian cartographic and cadastral authority IGAC (Licence CC BY 4.0)

the agro-export interest moved to another agribusiness hotspot on the Caribbean coast, some 500 km southwest, near the Gulf of Urabá. The demise of Magdalena's leadership in the banana-export market unleashed fierce competition for UFCO's lands. The company sold off most of its assets to Incora (the Colombian Institute for Agrarian Reform), which was intended to divide the farms among landless peasants and former banana workers (Bucheli 2005). In reality, Santa Marta elites and mid-size farmers with the necessary financial muscle and political connections managed to purchase or simply grab most of UFCO's land (Zamosc 1986, 42–3). When peasant collectives attempted to occupy the estates and required the fulfilment of agrarian reform promises, they were kicked out by the police or by groups of local gangs commissioned by powerful landowners (Britto 2020, 49–50). Throughout the 1970s and 1980s, some local landlords attempted to acquire state support in order to relaunch their banana-export activities, but most of the former UFCO estates had been turned into pastures dedicated to extensive cattle ranching (Bucheli 2005; Britto 2020, 50). Only in the late 1980s did the banana economy rise again in northern Magdalena (Viloria de la Hoz 2008). This time, as the next chapter will show, the boom brought about its share of social tensions and labour struggles. They would be solved at gunpoint by the new security providers – paramilitary groups.

The agribusiness north cannot be dissociated from two other neighbouring territories. The first is the Sierra Nevada mountain range to the east, which became a strategic point for the armed competition between rival paramilitary factions and the FARC guerrillas. From the 1970s, the Sierra became a hotspot for drug smuggling. Early in that decade, marijuana production transformed it into 'Colombia's first drug paradise' (Britto 2020). Marijuana's origins were not exogenous to the region's history, as it stood in the crosshairs of local power struggles and global capital flows. To fund, grow, transform, and smuggle the crop, a new breed of entrepreneurs relied on idle capital from the banana district and labour provided both by smallholder coffee growers in the Sierra and by unemployed agribusiness workers from the lowlands (Martínez and Renán 2011; Britto 2020). Marijuana was not just Colombia's first move into the global market for narcotics; it also put the country on the map for the global moral crusade aiming to suppress them. Following a joint initiative by Colombian and US officials, the first large-scale anti-drug operation in the history of the country was launched in November 1978. Over 6500 Colombian soldiers supported by undercover DEA agents were deployed in the Guajira peninsula and the Sierra Nevada mountains to crack down on the marijuana smuggling networks (Britto 2020, 174). While this – and later efforts – were successful at disorganizing the traffic, in the long run they did nothing but favour more violent competition among traffickers, which eventually gave rise to a new generation of smugglers specialized in the much more lucrative business of cocaine exports (Britto 2020, 201). As a matter of fact, the Sierra mountains provided excellent conditions for trafficking, as they were both geographically isolated and located at short distance from sources of chemical precursors (the Santa Marta port) and financial services (the cities of Santa Marta and Barranquilla). Moreover, as the northern slopes of the Sierra plunge into the sea, they result in coves and bays that create a suitable environment for embarking drugs (Betancourt and García 1994; Ardila *et al.* 2012). One of the major consequences of this history of drug trafficking and territorial competition is that it favoured the emergence of 'violence entrepreneurs', i.e. groups essentially focused on 'enabling organized force (or organized violence) to be converted into money or other valuable assets on a permanent basis' (Volkov 2002, 27). Initially established within the context of the violent and unregulated drug smuggling sector, these criminal groups gave birth to the first anti-communist militias in the 1980s.

South of the banana district and the Sierra mountains extends a region of fertile plains and savannahs dominated by cattle ranching. The scramble for land in this region is relatively recent. It was linked to the construction of the first railways in the 1930s, which sparked the interests of investors and hacienda owners, but also the migration of landless peasants, most of them looking for land on the fertile banks of the Magdalena river, especially beginning in the 1950s (Becerra and Rincón 2017). The centre and south of Magdalena was home to most of the land conflicts of the 1970 and 1980s, as organized peasants fought to gain access to land that was monopolized by large and often absentee landlords (Zamosc 1986). Peasants affiliated with the National Peasant Association (ANUC) faced coordinated attacks from landlords, police, and local gangs from the early 1970s. While no guerrilla activity would be reported in the department before the early 1980s, by the mid-1970s, local owning classes were demanding decisive state action against the peasants, under the pretence that ANUC collectives were infiltrated by the rebels (Becerra and Rincón 2017, 44–6). Beginning in the second half of the 1990s, paramilitary leaders saw in this long history of agrarian conflicts an appropriate bedrock for settling down and garnering the support of landed elites.

My focus on Magdalena will extend through this chapter and the next. I will also come back to the case study in the last chapter of the book, as I will focus on the legacies of paramilitary rule. In the following pages, the first section will describe the emergence of the first Magdalena militias, within a context characterized by drug booms, guerrilla expansion, and leftist mobilization. The second section will show how the local militias were sidelined and co-opted by an outsider group – the AUC – which was, however, closely linked to the interests of local power holders. I will then show how land grabbing was one of the various effects of paramilitary territorial and political power. The next chapter will then adopt a micro-view, focusing exclusively on the banana district and studying a set of land dispossession cases, demonstrating how they are embedded in the transformations of agrarian capitalism and global value chains.

Magdalena paramilitaries: from mobsters to anti-communism

Magdalena came onto the radar of violent conflict analysis in the 1980s, as a combination of guerrilla expansion, political leftist activism, and paramilitary repression translated into unprecedented levels of violence. This section will show that the extension of rebel activities to the region, posing a tangible threat to dominant classes, generated a demand for security that the state was unable to satisfy. Instead, this demand was met by existing violence entrepreneurs, whose origins were tied to the history of marijuana and cocaine smuggling. While these various groups strived to upgrade their firepower to the standards set by the rebels, they were largely unsuccessful at resisting guerrilla expansion. However, their most effective actions were directed against civil leftist activism, which was annihilated in the region between the late 1980s and early 1990s.

Armed escalation

In the early 1980s, the primary security threat for Magdalena's owning classes came in the form of guerrilla rebels, whose attempts to settle in the region dated back to the early 1980s. The FARC set up their first armed 'front' in the Sierra Nevada area in 1983 (Aguilera Peña 2014, 148). Their expansion into Magdalena was part of a larger strategy whose aim was to extend their influence throughout the northern regions of Colombia (Echandía 2006). The proximity of the agribusiness zone and the city of Santa Marta represented further opportunities for creating racketeering schemes. The thriving narcotics economy was also a driving force, as the group became increasingly involved in the first stages of the drug business. However, this involvement earned the rebels the enmity of many established smugglers. Intending to replicate the strategy that they had followed in southern Colombia, the FARC sought to become an intermediary between drug trafficking networks and the peasants who grew the marijuana and coca crops.[2] Their political discourse insisted on their will to protect these smallholders from abusive drug kingpins, but that did not prevent them from becoming abusive rulers themselves. The ELN's expansion shows a similar pattern. The group arrived in the department in the mid-1980s, attracted by the riches of the banana district and by the railway transporting coal from the Cesar department to the Santa Marta port. The rebels settled in the central plains of the province, where they extorted racket money from cattle breeders and other agricultural entrepreneurs (Carreño and Silva 2014). In this region characterized by violent land conflicts between large landlords and peasants, the group worked to entangle itself in existing antagonisms as a way of reaping the support of rural impoverished populations.

From the late 1980s, the territorial presence of the rebels began to be felt through a series of attacks on police stations and transportation infrastructure. By the early 1990s, the guerrilla had enough firepower to successfully ambush police or military troops, inflicting high death tolls. In 1991, 12 policemen were killed by the FARC while travelling between Santa Marta and Ciénaga. In 1993, seven soldiers from the Arhuacos battalion were killed in an ambush in the village of Parranda Seca. The rebels' territorial influence bolstered their capacity to milk the local owning classes and corporations. Racketeering in the agribusiness sector was an attractive source of income. This involved kidnapping plantation owners, assassinating farm managers, and burning down infrastructure.[3] According to political scientist Priscila Zúñiga, there were 75 murders and 250 arson cases linked to guerrilla racketeering in the agribusiness zone between 1986 and 2000 (Zúñiga 2007). In some cases, this affected the most prominent members of the business milieux. In 1996, two of the most important representatives of the agribusiness elite were kidnapped in the villages of Gran Vía and Riofrío, in the heart of the banana district: Fernando Campo Vives, who was an important entrepreneur and member of an influential family, and Rafael Pérez Dávila, former governor and president of the banana producers' association. In 1998, the FARC kidnapped Peter Kessler, general manager of the Colombian subsidiary of Dole, the US agrifood giant. A military operation was set up to release him, but he was killed in the crossfire.[4] Other business sectors were also targeted by guerrilla racketeering. The transportation and export of coal extracted from mines located in the neighbouring department of Cesar provided ample opportunities for extortion schemes, as a railway line of more than 250 kilometres is needed to reach the port from the coal mines. The FARC and the ELN reportedly began racketeering the US mining corporation Drummond around 1996 (Salinas Abdala *et al.* 2018). In the following years, the guerrillas bombed the railway, burned the company's trucks on various occasions, and were even able to raid the coal shipping facilities in the port.

The pervasiveness of the rebel threat opened a host of opportunities for a range of existing violence entrepreneurs, who were more than willing to offer protection services to agribusiness entrepreneurs and commercial establishments in the city. The history of these violent entrepreneurs is closely tied to the transformations in the criminal groups present in the region. The marijuana boom generated a large number of gangs who relied both on their capacity to use violence and their political connections with local elites to thrive. The best example of these characters is provided by Hernán Giraldo, a migrant from the Andean highlands who would later become the top paramilitary chief of Magdalena. Giraldo set up a small militia in the midst of the troubled times of the 1970s and staffed it with members of his extended family. He used it to control profitable transportation routes in the northern slopes of the Sierra, where marijuana and later cocaine was transported from the mountains to the natural ports on the Caribbean coast. In the 1980s, Giraldo extended his role as a protection contractor to certain neighbourhoods of Santa Marta. He operated a racketeering scheme in the city's market and invested his profits in the legal economy. The FARC's arrival in the Sierra mountains represented a threat for this militia, and led Giraldo to scale up his firepower in order to repel the rebels. By the end of the 1980s, Giraldo's militia had absorbed smaller groups and consolidated his grip over the northern slopes of the mountain. His growth was supported by profitable alliances with powerful cocaine smugglers from Medellín, as he was a critical actor in securing transportation routes and production facilities located in the Sierra mountains. This resulted in a violent rule over the population in his areas of influence, who he kept under strict surveillance in order to stave off any subversive activity. By the 1990s, he owned plots of some of the best land surrounding the city, as well as a supermarket chain and other commercial establishments such as billiard rooms and restaurants.[5] The Giraldo group is only the largest and longest-surviving example of a large class of violence entrepreneurs. Dozens of other gangs, such as the Rojas, the Cuquecos, the Meriños, or the Durán followed a similar, albeit less successful trajectory.

While undertaking illegal activities by definition, these violence entrepreneurs were able to set up legal facades and relationships with local law enforcement. One of the ways of obtaining a legal cover was through the creation of private security firms, which allowed them to legally purchase firearms and ammunition, as well as to officially bill their corporate clients. This possibility was enhanced by a 1994 law that turned security firms in zones of guerrilla activity into key collaborators for the military. Taking advantage of a very loose oversight procedure, these firms were able to purchase high-calibre weapons and other military-style equipment (Grajales 2017c). The legislation provided legal cover for the transportation of this kind of heavy weaponry, facilitating the expansion of the militias from their strongholds into new territories (Salinas Abdala *et al.* 2018, 80). In 1995, Giraldo utilized this new legal framework to create a shell company under the name of 'Conservar'.[6]

There is significant evidence of the links between Magdalena paramilitary groups and the army. Once again, the best-documented example is provided by Hernán Giraldo. As guerrilla activity escalated in northern Magdalena over the second half of the 1990s, Giraldo attempted to reinforce his group's combat capabilities. In 1996, with the support of several military officers posted at the Santa Marta base, he made 30 young men who he considered loyal to him enrol in the army. They were placed in the same platoon and sent out for joint operations in the Sierra mountains conducted in concert with Giraldo's forces. Once they had served their term, the young recruits were released and integrated into Giraldo's group. They had received far better training than regular militiamen and had embraced military rules and discipline, which they could infuse to the rest of the group.[7]

The violence entrepreneurs' territorial influence enabled them to receive compensation in exchange for ensuring electoral support. The militias' territorial control enabled them to provide votes to their allies. Giraldo, who had the largest social networks in the Sierra, was particularly well positioned to become an electoral broker. A peasant from a Sierra community remembers:

> Don Hernán said that we had to vote for this guy or that guy. We always listened to him, because Don Hernán knew what was best for the community. At the time, it was not yet out of fear that we obeyed, but out of respect, because we recognized in Don Hernán our 'natural leader' (*jefe natural*).[8]

Giraldo also funded political campaigns. This could provide the basis for nascent local political careers, as in the case of Héctor Rodríguez Acevedo, who was elected to Santa Marta's city council to represent the communities of the rural zone of Guachaca, at the heart of Giraldo's area of influence. But Giraldo's influence was also able to reach higher spheres. For instance, according to the Colombian Supreme Court, Enrique Caballero Aduén, a local patrician and former House representative, was elected to the Senate in 1994 and 1998 with Giraldo's financial and electoral support.[9]

A similar combination of political and military support is present in the case of other paramilitary militias, such as that of the Rojas. According to Adán Rojas, the patriarch of the family, and Rigoberto, his son and lieutenant, the expansion of his influence to the south-west of the Sierra and the banana district was supported by politicians, drug kingpins, and military commanders (Salinas Abdala *et al.* 2018, 71). Both men testified in court that these political links had been instrumental in connecting them with other paramilitary militias elsewhere in the country. They also testified that they led several joint raids with members of the Córdoba battalion, an army unit headquartered in Santa Marta (Carreño and Silva 2014, 105–6).

How efficient were these militias in fighting the guerrillas? In spite of these broad networks of support, they were not capable of protecting their clients against the rebel threat. The 1990s saw an average of 23 cases of kidnapping per year. Beginning in 1998, as the ELN and the FARC began to deploy a much more aggressive kidnapping-for-ransom strategy across Colombia, the number of cases soared, peaking at 119 cases in 2002 (Policía Nacional de Colombia 2008, 263–4). Guerrillas were generally able to overpower most militias. In the early 1990s, the Rojas suffered a series of guerrilla attacks in their fiefdom of Palmor, and were only saved from complete annihilation by Giraldo's temporary support.[10] Smaller groups were confined to the relative safety of urban areas, unable to compete with the guerrilla groups' territorial influence in rural areas. Except for Giraldo, all the other armed groups that originated in the context of the 1980s were either absorbed by larger players or remained loose networks of killers, nothing close to an anti-communist army. They were often enmeshed in violent rivalries for territory and influence, as illustrated by the brutal opposition between the Giraldo and Rojas group, who went from friends to foes in 2000, which entailed the disarticulation of the latter.[11] However, this mild success at responding to their purported objective (countering the guerrilla) ignores the fact that 'protection' and 'order' are malleable notions (Lévy-Aksu 2012). In Magdalena, this did not exclusively mean protection from guerrilla activity, but equally – and sometimes primarily – the repression of social and political mobilizations that threatened the dominant classes' stronghold over local institutions.

Political violence

In addition to the rebel menace, and perhaps equally threatening to the supremacy of local rulers, was the growth of leftist opposition. This sort of activism drew its origins from a series of 1970s grassroots mobilizations, which were emboldened in the 1980s by a national context favourable to political participation. The peace-process with the FARC had bolstered the electoral mobilization of the left across the country, and had brought together a large front of social and political organizations under the umbrella of the party Unión Patriótica (UP). Other leftist groups, while they did not join the UP, took advantage of this favourable climate to seize electoral victories, often at the local level. In Magdalena, Marcos Sánchez Castellón embodied this young generation of leaders. A former student leader and grassroots activist, Sánchez was the candidate of a broad leftist front in the 1988 mayoral race in Santa Marta.[12] Sánchez demonstrated significant skill in mobilizing the populace. In early August 1987, the young lawyer organized a demonstration to protest the high prices charged by the local electricity supplier. During the march, hundreds of people burned their electricity bills in the cathedral square. This was a popular mobilization over bread and butter issues, but its significance was highly destabilizing in a political arena characterized by patrician supremacy. The act was interpreted at the time as a challenge to the families who held municipal power.

Two days after the march, Sánchez became the first victim of a new breed of political violence, as he was assassinated while leaving a friendly meeting with labour union activists.[13] When questioned by Radio Galeón, the Magdalena police chief, Colonel Pedro Antonio Herrera, justified the homicide by accusing the victim of being a member of the FARC.[14] The day after Marcos' death, a leaflet circulated in the city taking credit for the assassination on behalf of a paramilitary group and accusing him of having been involved with the guerrilla groups. Later, the local police claimed that the killers came from the Giraldo group, but this information has never been confirmed.

The following years saw the purposeful elimination of the left from Santa Marta and northern Magdalena. A few weeks after Sánchez's death, Adalberto Pertuz Bolaño, Sánchez's campaign manager, was murdered in the city. The Sánchez and Pertuz murders opened a cycle of violence that had methodically gone about destroying the local UP chapter within a few short years. Other local Magdalena chapters were made to endure a similar brand of violence. In the town of Aracataca, where the UP was allied with the Liberal Party and participated in the municipal government, the party's local leader, Juan Alberto Uribe Meléndez, was murdered on 30 June 1990. The repression against the left did not end with the disappearance of the UP from the political landscape. A new party, the Alianza Democrática M-19 (ADM-19) was born in 1990 after the M-19 guerrilla group signed a peace agreement with the government. Its main figure in Santa Marta was Ricardo Villa. Villa had been elected to the Senate in 1986 as a member of the Liberal Party. He joined ADM-19 from its inception and launched a local political platform focused on fighting corruption. He was murdered on 23 December 1992.

Paramilitary violence reshaped the entire political playing field. For political activists, the risk of being murdered eventually came to determine what a feasible political strategy was and the kind of alliances that needed to be made to obtain protection. Some of the members of this generation left the region or abandoned politics, but many simply moderated their words and their expectations in order to save their lives. This allowed a certain political left to survive and to rise again in the late 2000s, when new spaces of power were opened. The victory of the left in the 2011 mayoral race in Santa Marta (and later elections), as well as in the 2019 gubernatorial race, are without a doubt rooted in these social networks. Some of today's victorious activists were the men and women I interviewed years before about their dead comrades. For many, this represents a triumph over past hardships. Yet, throughout the 1990s and the early 21st century, the political movements that originated in the vitality of the social struggles of the 1970s and 1980s were kept at the margins of political power. While an outsider like Marcos Sánchez threatened the continuity of the political power structure in 1988, violence disintegrated these political networks and established insurmountable barriers to entry in Colombian politics for three decades.

In newspaper reports, most of these assassinations were attributed to so-called 'dark forces' (*fuerzas oscuras*), a vague term designating the fact that state agents or political elites were believed to be behind the murders (García-Peña Jaramillo 2005). Today, thanks to the testimonies of former paramilitaries, we know that this label actually concealed the alliances between local politicians, security personnel, and various groups of violence entrepreneurs. However, most of this testimony shrouds the true motives of many of these killings in a haze of anti-communist rhetoric. When questioned by judges and prosecutors, former paramilitary leaders systematically accuse their victims of belonging to leftist movements and being associated with guerrilla networks. Confronted by his judges about the killing of two professors and a University of Magdalena student, paramilitary chief Hernán Giraldo simply stated that it was because they were Communist Party members and concluded: 'I told my men that anything that smelled like guerrillas had to be eradicated' (Medina 2007).

These accusations often conceal other, more prosaic motives. The case of Ricardo Villa provides a good example of this. While Giraldo admitted having ordered Villa's killing at the request of politician and drug kingpin Jorge Gnecco, he claimed that the assassination was linked to Villa's leftist militancy. A closer look at the case reveals the true motives of the crime. Since the late 1980s, the Gnecco family had been involved in the flourishing real estate sector. One of the key geographical areas for the construction business in Santa Marta was Pozos Colorados Bay, most of which was public domain land. According to information compiled by Villa, Jorge Gnecco had set up a complex operation to grab these lands. He reportedly sent poor families to occupy the land and paid lawyers to assist them in legal proceedings. They would obtain a property title granted by the state. However, as soon as the title deeds were established, the land was purchased by Gnecco's frontmen and became available for real estate projects. Ricardo Villa had investigated this case because he represented the legal interests of the national tourism corporation, one of the main public agencies affected by the occupation of the bay. In his national newspaper columns, he denounced Gnecco's scheme.

The history of Magdalena's violence entrepreneurs, as elsewhere in the country, is one of transformation of criminal gangs into political actors (Gutiérrez 2004). While most of these groups' origins are linked to the marijuana economy, several of them made an anti-communist turn around the mid-1980s. This was the result of their own conflicts with guerrilla groups, but also of a demand for protection from the region's elites. However, this demand was not focused solely on the risks represented by the rebels themselves. Leftist movements and social mobilization were, from the point of view of many among the region's elites, equally pressing matters. As such, the presence of these violence entrepreneurs became a resource for buttressing ruling positions. More than agents in the counterinsurgency efforts that raged at the time, they acted as guardians of an established social order.

Building a monopoly of violence

By the mid-1990s, Magdalena, as many other parts of Colombia, was a highly polarized and violent social setting. However, no single armed actor dominated the region. The state could claim nothing close to a monopoly of violence. The guerrillas were influential, but they faced armed resistance from the military and paramilitary groups, several of which were competing in the violence entrepreneurship market. This would change with the arrival of outside militias, members of the AUC confederacy, who took control of the whole region between the late 1990s and the early 2000s. The arrival of this much more powerful actor transformed relations between paramilitary groups and local elites. The former were no longer simply security providers for the latter, but were playing their own game. They were pursuing a project of political and economic domination, which included the elimination of dissident voices and tight control over social life. However, this project was by no means opposed to the interests of the dominant classes. Most politicians already in power managed to adapt to, and often to take advantage of, this new setting.

The AUC's expansion and rule

Salvatore Mancuso, who would eventually become the most powerful figure in the national paramilitary confederation, was in charge of setting up a group in Magdalena in 1996. According to him, at its inception this group had only 40 men, but it grew at a rapid pace. By 1999, it operated both in Magdalena and in neighbouring Cesar, and it had 300 men at its orders.[15] From that year, the central role in paramilitary operations for the entire northern coast was played by Rodrigo Tovar Pupo, aka Jorge Cuarenta. Cuarenta had an elite background, as he came from a powerful family of landowners from Valledupar, the capital city of Cesar. This allowed him to play an instrumental role in introducing the paramilitaries to the ruling families of the region (Dudley 2009). He had also extensive ties to military officials, and played a critical role in strengthening the ties between paramilitary groups and the army (Salinas Abdala *et al.* 2018, 79). Cuarenta settled his headquarters in central Magdalena, in Sabanas de San Ángel, a region of extensive pastures where land is dedicated to cattle breeding and land conflicts have been pervasive since the 1970s. The arrival of these newcomers resulted in sharp increases in the amount of violence, as the number of victims of mass killings illustrates. Eleven people were murdered in massacres in 1998, 42 in 1999, and 103 in 2000 (ODH 2014). Massacres, which are defined by Colombian law enforcement as the collective murder of four or more people, are a reliable benchmark for measuring paramilitary activity; these egregious acts were massively perpetrated by paramilitary militias as a way of instilling fear and causing the forced displacement of whole communities (CNMH 2016).[16] This is particularly true in places that became paramilitary strongholds such as Magdalena.

FIGURE 2.1 Massacres in Magdalena. Source: the author, data from ODH (2014)

The motivations for paramilitary expansion remained diverse, as they combined political, economic, and territorial aspirations. The expansion was a way of presenting the AUC as a truly national group capable of having a similar reach to that of guerrilla rebels (Echandía 2006; Romero 2007). The territorial reach also responded to economic motives, as it gave access to new sources of funding, such as protection money from companies operating in the region, public budgets from local governments within their sphere of influence, and profitable cocaine smuggling routes (Valencia 2009; Romero 2011). AUC leaders stressed other justifications. At the core of Cuarenta's justification for his actions stood the terror that kidnapping generated among the owning classes.

Claiming to fight in the defence of liberty and private property was a bold legitimizing discourse. This could entail staging high-intensity violence as retaliation for prominent cases of kidnapping. The most telling example of this is the paramilitary reaction to the abduction of nine members of the hunting and fishing club of Barranquilla, including one town councillor, by the ELN in June 1999. The kidnapping took place as the victims were boating in the Ciénaga Grande wetlands, a popular hobby for members of the local upper classes. A few months later, the paramilitaries unleashed an egregious punitive response against the fisherfolk communities of the wetlands, accused of having provided logistical and informational support to abet the kidnapping (Carreño and Silva 2014; Sarmiento 2016). Between January and November 2000, paramilitary militias murdered at least 50 people. These killings were more than an attempt to suppress the ELN's networks in the area. They were a message sent to those terrified by the risks of abduction that protection was being offered.

But for the AUC paramilitaries to become prominent actors in the region, it was not enough to attack the communities who were supposedly favourable to the rebels' cause. It would also, and even primarily, be necessary to absorb other groups of local violence entrepreneurs – whether they wanted to be absorbed or not – into their organization. Small gangs in the banana district were made a part of the AUC. In southern Magdalena, the group founded by Chepe Barrera, a rancher from the area, integrated Cuarenta's army. However, the greatest challenge to the AUC's hold on the area was Giraldo's group, which at the turn of the century was the only independent militia in the region. Overpowering him was both an economic priority, as he controlled the most profitable route for cocaine smuggling, and a political milestone in the constitution of a national paramilitary confederation. The AUC's forced takeover of Giraldo's organization required the top confederacy commander, Carlos Castaño, to deploy over 400 men in November 2001. The local militias did not have the means to contain such a firepower. The operation was led by Rigoberto Rojas, Adán's son, who knew the region and had a deep hatred for Giraldo. In parallel to the military offensive, Castaño's hitmen assassinated several Santa Marta businessmen linked to Giraldo, including some of his frontmen. Cornered in the heights of the Sierra mountains, Giraldo grudgingly agreed to put himself under the command of Jorge Cuarenta. The military control of the Sierra was entrusted to Rigoberto Rojas, now an AUC member. In spite of this reversal of roles, Giraldo remained the 'political chief', i.e. the man in charge of relations with the Sierra communities and the political elites of Santa Marta.

The takeover of the Giraldo militia by the AUC was followed by an offensive launched against guerrilla positions in the Sierra mountains. The paramilitary group's objective was to extend its control over the whole northern and eastern slopes; especially in the upper heights, where the rebels occupied entrenched positions.[17] These efforts were largely successful, as they corresponded to counterinsurgent operations deployed by the Colombian military. By the end of 2003, the parallel deployment of military and paramilitary operations had isolated the FARC and the ELN in the most remote areas of the mountain. In the case of the former, external attacks seem to have fallen at a time of internal crisis, as the commander nominated in 2001 by FARC's central secretariat at the head of Magdalena's troops had alienated both his own authority and the guerrilla's support among civilians due to his conspicuous lifestyle (Aguilera Peña 2013, 60). While the guerrilla group did not fully disappear, its capacity to operate in the urban zones and the banana district, as well as to reach transportation routes, was severely curtailed.

Political influence

By 2002, the AUC were the uncontested armed actor in the Magdalena lowlands. One of the most significant consequences of this control was the capacity to gain an unprecedented level of political influence. The pursuit of political clout by violence entrepreneurs was not a new phenomenon, as Giraldo had endeavoured to achieve a certain degree of influence over elected officials. Funding political campaigns and mobilizing the votes of the people living under their influence was a way of making political allies who could defend his interests. However, the extension of the AUC's control over Magdalena's territory transformed this dynamic. Between the late 1990s and the early 2000s, Cuarenta did not simply reproduce the old strategy of buying political support. He strived to transform the militias into key regulators of local political rule.

The transformations in the selection of political officials can be illustrated by the trajectory of former governor Trino Luna. Luna was an insider. His mother had been a congresswoman and he was acquainted through marriage with the Dangonds, a powerful family of agribusiness entrepreneurs from Ciénaga. However, Luna's political capital was mainly limited to the south of the province, and he was seen by the patrician families of the north as lacking social pedigree. To overcome this impediment, Luna entered an alliance with Jorge Cuarenta. His relationship to the paramilitaries was close-knit, as his own brother, Juan Carlos, was one of Cuarenta's own lieutenants. Even more importantly, Luna became a broker between the political barons of small municipalities and the paramilitaries. In the words of a former politician, he 'was the paramilitaries' man with rural mayors, and the mayor's candidate with the paramilitaries'.[18] This position enabled him to overpower the Santa Marta patricians and to be nominated as the paramilitaries' candidate for the 2003 polls. Cuarenta prohibited any other candidacy, which led Luna to be elected with 81 per cent of the vote, with a blank vote as his only competitor. Those who dared to denounce the scam were severely punished, as it happened with former mayor of El Banco Fernando Pisciotti, who filed a complaint in 2003 on the paramilitaries' political influence and was murdered in December of that year. In 2011, Trino Luna's brother was indicted for this crime, but the investigation was stalled until his death by natural causes in 2016.

The strategy to seize political control over congressional elections was more complex. The paramilitary forces strived to maximize their electoral potential by mobilizing the entire vote of a single municipality in favour of the same candidate. As such, voting in towns under paramilitary influence was massively concentrated. Paramilitary allies would gather 90 per cent of the ballot in Concordia and 95 per cent in El Piñón, Pedraza, and Tenerife. The strategy was successful, and Cuarenta managed to elect three of his allies to the Senate, and the same number to the House. In spite of its pervasiveness, this strategy to gain influence did not result in a notable transformation in the composition of the legislature. While some politicians were excluded by their opposition to the paramilitaries, most of the congresspeople elected in 2002 and 2006 had previous political careers. I ran a biographical survey on 30 candidates elected in four Caribbean departments (Cesar, Córdoba, Magdalena, and Sucre) who were targeted by judicial inquiries on their ties with paramilitary groups (Grajales 2017b, 76). Only six of them lacked significant political experience prior to the early 2000s. The few novel figures were always supported by local patricians.

Paramilitary groups' friends in Congress were instrumental in securing the nomination of people on their payroll at the head of key regional public agencies. Once the management of local public agencies and offices was secured, their budget could be siphoned by the alliance of politicians, bureaucrats, and paramilitary commanders

(Romero 2011). One of the first cases documented by scholars and journalists concerned hospitals and public health agencies. Paramilitary groups levied a tax on the contracts passed by these institutions. In some cases, bogus contracts were signed with shell companies controlled by the paramilitaries themselves or their associates (Romero *et al.* 2011). Carlos Tijeras, who was Cuarenta's lieutenant in charge of the Magdalena's banana district, testified in court that all important nominations in the health sector were submitted to his approval. When necessary, violence was used to enforce his rule. As a matter of fact, the imprisoned paramilitary acknowledged having ordered the murder of the Zona Bananera hospital director, who had supposedly voiced opposition to the embezzlement scheme.[19] Violence could also target ordinary workers with access to sensitive information. Zully Esther Codina, an accounting officer at the Santa Marta hospital who was also a trade union leader and a journalist for a local radio, was murdered in 2003.[20]

Coercion, institutional power, and accumulation

The occupation of spaces of power was also used to violently seize and accumulate land in the paramilitary groups' areas of influence. While land grabbing required the use of coercion, the control of local institutions was instrumental to turning these ill-gotten assets into legally recognized property (Grajales 2011; Ballvé 2012; Peña-Huertas *et al.* 2017). In Magdalena, as elsewhere in Colombia, land was vacated through direct violence and threats. Violence against peasant communities could be staged in a brutal manner in order to cause displacement. In central Magdalena, paramilitary attacks on peasant communities were the continuation of a long history of conflict concerning land. A wave of paramilitary violence began in 1996, with the arrival in central Magdalena of ACCU militias. Among their first victims were some of the historical peasant leaders of the area, people like Corina Barón and Andrés Vallejo, who were murdered by Jorge Cuarenta's men (Becerra and Rincón 2017, 69–70). These murders were instrumental in dismantling the peasant movement, but they also fuelled high levels of forced displacement. As an illustration, in the municipality of Pivijay, which was characterized by a long history of peasant mobilization and landlord-sponsored violence, 5397 victims of forced displacement were registered in 1999 (ODH 2014).

Violence was a necessary but insufficient ingredient of dispossession. Control over local institutions was critical to securing ownership rights for paramilitary leaders themselves, their frontmen, and their business partners. The recording of land deals exemplifies this point. Recording a land deal in Colombia involves two institutions: notaries and registration offices (*oficinas de registro de instrumentos públicos*). Notaries are public officials placed under the surveillance of a national body, the SNR (*Superintendencia de notariado y registro*). Until a recent reform, most were nominated by decree, primarily through the support of a powerful political sponsor (often a congressperson). Notaries are in charge of certifying the legality and regularity of a transaction. When this has been done, the registration office records the acquisition of the property and updates the registrar with the name of the new owner. Transformations in the ownership of a plot, such as splitting it into several parcels or merging various plots into a single estate, must be recorded at the registration office. Registration officers are public servants and are considered as SNR staff. However, even if their nomination is an internal decision of the SNR, the agency consults with local officials – often governors – before making a new nomination.

Paramilitary commanders relied on their political allies to secure 'safe' nominations to these positions, i.e. to place people who would participate in their schemes. The role of notaries in land dispossession in Magdalena has been very finely documented by Quinche Roa *et al.* (2018); according to these authors, after landowners would receive threats from paramilitaries forcing them to sell their plots, the purchaser would pay them a visit in company of the notary, who would then proceed to record the transaction. In other cases, notaries would deliver false powers of attorney to land brokers working with the paramilitaries, who would then sign a transaction on behalf of the real owner. Registration officers were instrumental in the steps that came after, as they provided the means to conceal the spurious origins of an estate. The most common strategy involved splitting and merging plots. After legalizing a purchase, a common element of dispossession was to merge neighbouring plots into a single estate (Quinche Roa *et al.* 2018). The new estate would have a clean file from which previous owners of the smaller plots were absent. In those conditions, proving that a plot had been violently grabbed required tracing back the history of the estate prior to its merging. As until very recently, most of the registration offices in rural areas kept their records on paper, this simple procedure could erase the history of a land plot (Gutiérrez and Vargas 2016).

Paramilitary land grabbing was rarely random; on the contrary, it was highly determined by the militias' social embeddedness in an owning class that had long resisted any attempts at land redistribution and had often resorted to violence to defend their privileges. Gutiérrez and Vargas (2017) accurately remark that 50 per cent of AUC commanders were cattle ranchers at the time they entered the militia. In Magdalena, several paramilitary figures came from the families of large landowners and livestock breeders. Jorge Cuarenta himself was a rancher, as were Chepe Barrera, a paramilitary commander from the south of the department, and Saúl Severini, one of Cuarenta's lieutenants in the central lowlands. One of the consequences of this social embeddedness is that paramilitary activities were, for many of these men, simply the continuation of old agrarian feuds with the involvement of more armed muscle. A good illustration of this is the violence unleashed against the peasant communities that had benefitted from agrarian reform land distribution policy. Thanks to the work of various civil society organizations (NGOs and law clinics), the cases of the peasant communities of Chibolo, a municipality located in the centre of the department, are among the best documented in Magdalena (see Becerra and Oyaga 2011; CJYC 2016). Chibolo was home to very high levels of land conflicts in the 1970s, as ANUC collectives fought to force the state to meet its land reform promises. Communities such as those of La Pola, La Palizúa, and El Encanto, managed to resist police, military, and landlord harassment and to force the government to intervene. They occupied idle estates belonging to some of the region's richest landowners and reclaimed the land for themselves. In the classic pattern outlined in the previous chapter, their aim was to solicit the involvement of Incora, which would expropriate or purchase the land they had occupied and then allot it to the peasant collective. While the expropriation procedures never succeeded, by the early 1990s, Incora had agreed to support the sale of the estates under a subsidized purchasing policy. By 1996, when the paramilitaries began harassing the peasants, the land titling operation was underway. The peasant communities were forcibly expelled between 1996 and 1999. Several peasant leaders and other prominent members of the community were assassinated as a way of infusing fear and shattering any form of resistance. Once the communities had been dispossessed of the land, it was occupied by the militias.

The paramilitaries' institutional networks were then put to work in the service of this land grabbing scheme. In 2011, the attorney general's office arrested five Incora officials, including the Magdalena office director and his legal advisor. The director, José Fernando Mercado Polo, signed a plea deal and confessed to having played an instrumental role in land dispossession. During his time at the head of the office between 2001 and 2003, he overturned the land allocation decisions for 134 estates, including 47 plots in Chibolo (Bonilla 2012, 156). In these cases, the Incora office had issued an administrative act recording the abandonment of the plots, obviously without noting that their occupants had been forcibly displaced (Becerra and Oyaga 2011). As abandonment was a legal cause for extinguishing the allocation, the plots could be allocated to a new owner. This pattern was repeated in other locations of central Magdalena, where a similar sequence of events transformed mobilized communities into dispossessed internal refugees.

Who were the beneficiaries of these land grabbing schemes? Our knowledge remains extremely partial and is contingent on judicial sources that have merely scratched the surface of the problem of dispossession. However, this partial knowledge provides a picture of the paramilitaries' extended political networks. In the cases documented by Quinche Roa *et al.* (2018, 27–34), the beneficiaries of land grabbing were Cuarenta himself, his own frontmen, and business associates, as well as, in some cases, paramilitary lieutenants and even some rank-and-file militia members. A typical example of a paramilitary business associate can be found in the case of Augusto Castro, a wealthy rancher from central Magdalena who accumulated thousands of hectares of land violently grabbed by the paramilitaries. Castro came from a patrician family; his sister had been mayor of his home town, one of his brothers was a member of Magdalena's legislative assembly, and the other was a senator (Verdad Abierta 2013). Either through the allocation scheme run by Incora, or with the support of notaries and frontmen, Castro accumulated land in Chibolo and the neighbouring municipalities of Ariguaní, Sabanas de San Ángel, and Plato (Quinche Roa *et al.* 2018, 42). Criminal investigations against Castro have shown that his political and institutional networks were helpful not only in legalizing land grabbing, but also in enabling him to reap further financial benefits. As a matter of fact, thanks to his brother's political support, he obtained subsidies from Corpomagdalena, the regional environmental agency, as well as a contract for timber production with the French multinational ONFI, in 2006.

These examples support the view that violent land grabbing was not a radical rupture in the social order, but the continuation of less radical forms of domination with bolstered resources. The development of paramilitary militias in Magdalena, as elsewhere in the Caribbean lowlands, did not upend existing forms of domination. Quite the opposite, their embeddedness in local histories of class domination determined their position as guardians of an exclusionary political arena and a proprietarian social order. These ideas will be at the core of the next chapter.

Conclusion

This chapter has presented the social mechanisms though which violence entrepreneurs established themselves as key actors in the defence of an exclusionary social order. I first traced the emergence of violence entrepreneurship as a phenomenon linked to the drug trafficking economy, which became pervasive in many Colombian regions from the late 1970s. Magdalena epitomizes a phenomenon observed nationwide: the interaction between groups of violence entrepreneurs and the local elites, spurred by the multiple threats faced by dominant classes during the last decades of the century (Romero 2003; Gutiérrez 2019). The relationships established between these two types of actors were determined by a shared interest: the reaction against both the threat from armed guerrillas and the spectre of the political left. While relations between leftist activists and professional revolutionaries were far from straightforward, the two categories were conflated into a single threat both by the discourse of governing elites and by paramilitary violence. The result was bloodshed of epic proportions, which proved only mildly effective against the guerrillas but managed to annihilate the political left, in Magdalena as elsewhere in Colombia.

In the second section, I moved on to describe the transformation of Magdalena's political landscape in the early years of the 21st century, as the grip of one armed actor – the AUC – became pervasive. Paying attention to both the economic and the political fields, I showed that the conquest to achieve territorial superiority enabled militia leaders to dictate the terms of political power, but also to secure the conditions of accumulation. Narrowing this view to the militias' participation to land grabbing, I argued that Magdalena is illustrative of a larger pattern of land dispossession, in which coercive power and institutional support nurtured a process of accumulation.

The next chapter will use this observation as a point of departure. The understanding of paramilitaries' political and territorial influence is a prerequisite for studying the economic role they played, often as agents of repression and land accumulation. However, the following pages will not be centred upon the militia groups' strategies, but on rural people's experiences of land struggles, dispossession, and resistance. My aim will be less to describe the intricate methods of dispossession – an already well-documented endeavour – but to analyse the embeddedness of dispossession in the political economy of agrarian capitalism.

Notes

1 In a similar way to the analysis on subnational authoritarianism developed by Gibson (2005).
2 Tribunal Superior de Bogotá, José Gregorio Mangones Lugo 2007-82791, 31 July 2015, p. 562.
3 I traced the escalation of guerrilla activity using the data on armed conflict furnished by the CINEP, accessed in 2009 and 2011. A more detailed account is available in Grajales 2017b, 33–4.
4 Tribunal Superior de Bogotá, José Gregorio Mangones Lugo, decision No. 2007-82791, 31 July 2015, p. 572.
5 Tribunal Superior de Barranquilla, Jancy Antonio Novoa Peñaranda, decision No. 2011-83374, 21 October 2014; Tribunal Superior de Bogotá, José Gregorio Mangones Lugo, decision No. 2007-82791, 31 July 2015; Tribunal Superior de Barranquilla, Hernán Giraldo et al., decision No. 2013-80003, 18 December 2018.
6 Tribunal Superior de Barranquilla, Jancy Antonio Novoa Peñaranda, decision No. 2011-83374, 21 October 2014, p. 42.
7 Tribunal Superior de Barranquilla, Hernán Giraldo et al., decision No. 2013-80003, 18 December 2018, p. 108.
8 Interview, Santa Marta, March 2009.
9 Corte Suprema de Justicia, Enrique Rafael Caballero Aduén, decision No. 33416, 9 March 2011.
10 Tribunal Superior de Barranquilla, Jancy Antonio Novoa Peñaranda, decision No. 2011-83374, 21 October 2014, p. 38.
11 Tribunal Superior de Bogotá, José Gregorio Mangones Lugo, decision No. 2007-82791, 31 July 2015, p. 585.
12 This was the first popular election for mayors, who were appointed by the governor until a 1986 reform.
13 I have traced the history of violence against leftist activist in the region mainly through oral sources, interviewing former activists and members of the victim's families. Few scholarly works on this topic exist. My analysis also relies on the works of Renán (2007) and Zúñiga (2007).
14 Letter from Robert Raven, from the American Bar Center to Virgilio Barco, President of Colombia, regarding the murder of Marcos Sánchez Castellón. 14 September 1988. National Security Archive.
15 Tribunal Superior de Barranquilla, Rolando René Garavito, decision No. 2011-83724, 11 July 2016, p. 52.
16 Some databases, such as the 'Noche y Niebla' project run by CINEP (2004) have endeavoured to document the murder cases that can be clearly attributed to paramilitary groups. However, while these data are extremely useful to study individual cases, it is much more difficult to draw quantitative conclusions, as the identification of a murder's perpetrator poses intractable difficulties in a context of pervasive violence such as Colombia's.
17 Tribunal Superior de Barranquilla, Jancy Antonio Novoa Peñaranda, decision No. 2011-83374, 21 October 2014, particularly p. 54–5.

18 Interview, Santa Marta, June 2011.
19 Tribunal Superior de Bogotá, José Gregorio Mangones Lugo, decision No. 2007-82791, 31 July 2015.
20 Tribunal Superior de Barranquilla, Hernán Giraldo *et al.*, decision No. 2013-80003, 18 December 2018, p. 720.

References

Aguilera Peña, M. 2013. Claves y distorsiones del régimen disciplinario guerrillero. *Análisis político*, 26(78), 45–62.
Aguilera Peña, M. 2014. *Guerrilla y población civil: trayectoria de las FARC 1949–2013.* Bogotá: Centro Nacional de Memoria Histórica.
Ardila, E., A. Acevedo, L. Martínez, and F. Silva. 2012. *Memorias de violencia: bonanza marimbera en la ciudad de Santa Marta durante las décadas del setenta al ochenta.* Santa Marta: Oraloteca. Universidad del Magdalena.
Avilés, W. 2007. *Global Capitalism, Democracy, and Civil-Military Relations in Colombia.* Albany, NY: SUNY Press.
Ballvé, T. 2012. Everyday State Formation: Territory, Decentralization, and the Narco Landgrab in Colombia. *Environment and Planning D: Society and Space*, 30(4), 603–22.
Ballvé, T. 2020. *The Frontier Effect: State Formation and Violence in Colombia.* Ithaca, NY: Cornell University Press.
Becerra, C. and F. Oyaga. 2011. *Desplazamiento forzado y despojo de tierras en Chibolo.* Bogotá: ILSA.
Becerra, C.A. and J.J. Rincón, eds. 2017. *Campesinos de tierra y agua: Campesinado en el Magdalena.* Bogotá: Centro Nacional de Memoria Histórica.
Betancourt, D. and M. García. 1994. *Contrabandistas, marimberos y mafiosos.* Bogotá: Tercer Mundo.
Bonilla, L. 2012. Élites, órdenes rurales y restitución de tierras en la costa Caribe. *Arcanos*, (17), 144–66.
Botero, F. and Á. Guzmán Barney. 1977. El enclave agrícola en la zona bananera de Santa Marta. *Cuadernos Colombianos*, 3(11), 309–89.
Britto, L. 2020. *Marijuana Boom: The Rise and Fall of Colombia's First Drug Paradise.* Oakland, CA: University of California Press.
Bucheli, M. 2005. *Bananas and Business: The United Fruit Company in Colombia, 1899-2000.* New York, NY: NYU Press.
Carreño, D. and F. Silva. 2014. *Los pueblos palafitos: Ese día la violencia llegó en canoa. Memorias de un retorno: Caso de las poblaciones palafíticas del complejo lagunar Ciénaga Grande de Santa Marta.* Bogotá: Centro Nacional de Memoria Histórica.
CINEP, (Centro de Investigación y Educación Popular). 2004. *Deuda con la humanidad, paramilitarismo de Estado en Colombia.* Bogotá: CINEP.
CJYC, (Corporación jurídica Yira Castro). 2016. *"Ya supimos por dónde vino el agua al coco": Relato de resistencia al despojo y análisis del proceso de restitución de tierras en Magdalena.* Bogotá: CJYC.
CNMH, (Centro Nacional de Memoria Histórica). 2016. *Basta ya. Colombia: memorias de guerra y dignidad.* Bogotá: Centro Nacional de Memoria Histórica.
DANE, (Departamento Administrativo Nacional de Estadística). 2016. *Tercer censo nacional agropecuario.* Bogotá: DANE.
Dudley, S. 2009. Entrevista a Rodrigo Tovar en la Correccional Washington DC. *InSight Crime*, 3 Jun.
Duncan, G. 2005. *Los señores de la guerra.* Bogotá: Planeta.

Echandía, C. 2006. *Dos décadas de escalamiento del conflicto armado en Colombia. 1986-2006.* Bogotá: Universidad Externado de Colombia.

García-Peña Jaramillo, D. 2005. La relación del Estado colombiano con el fenómeno paramilitar: por el esclarecimiento histórico. *Análisis político*, (53), 58–76.

Gibson, E. 2005. Boundary Control. Subnational Authoritarianism in Democratic Countries. *World Politics*, 58(1), 101–32.

Gill, L. 2016. *A Century of Violence in a Red City: Popular Struggle, Counterinsurgency, and Human Rights in Colombia.* Durham, NC: Duke University Press.

Grajales, J. 2011. The Rifle and the Title: Paramilitary Violence, Land Grab and Land Control in Colombia. *The Journal of Peasant Studies*, 38(4), 771–92.

Grajales, J. 2017a. Privatisation et fragmentation de la violence en Colombie: l'État au centre du jeu. *Revue française de science politique*, 67(2), 329–48.

Grajales, J. 2017b. *Gobernar en medio de la violencia. Estado y paramilitarismo en Colombia.* Bogotá: Editorial Universidad del Rosario.

Grajales, J. 2017c. Private Security and Paramilitarism in Colombia: Governing in the Midst of Violence. *Journal of Politics in Latin America*, 9(3), 27–48.

Gutiérrez, F. 2004. Criminal Rebels? A Discussion of Civil War and Criminality from the Colombian Experience. *Politics & Society*, 32(2), 257–85.

Gutiérrez, F. 2019. *Clientelistic Warfare: Paramilitaries and the State in Colombia.* Oxford: Peter Lang.

Gutiérrez, F. and J. Vargas. 2016. *El despojo paramilitar y su variación: quiénes, cómo, por qué.* Bogotá: Universidad del Rosario.

Gutiérrez, F. and J. Vargas. 2017. Agrarian Elite Participation in Colombia's Civil War. *Journal of Agrarian Change*, 17(4), 739–48.

Hristov, J. 2014. *Paramilitarism and Neoliberalism: Violent Systems of Capital Accumulation in Colombia and Beyond.* London: Pluto Press.

LeGrand, C. 2012. Living in Macondo: Economy and Culture in a United Fruit Company Banana Enclave in Colombia. *In*: G.M. Joseph, C. LeGrand, and R.D. Salvatore, eds. *Close Encounters of Empire.* Durham, NC: Duke University Press, pp. 333–68.

Lévy-Aksu, N. 2012. *Ordre et désordres dans l'Istanbul ottomane (1879-1909).* Paris: Karthala.

Martínez, L. and W. Renán. 2011. Memorias y narrativas: una reconstrucción del contexto de violencia en el Magdalena (1980-2009). *In*: F. Silva, ed. *Reconstrucción de la memoria oral de los desmovilizados y desplazados en los departamentos del Magdalena, Cesar y Guajira entre 1980 y el 2009.* Santa Marta: Oraloteca. Universidad del Magdalena, pp. 358–531.

Medina, S. 2007. Hernán Giraldo reconoció crímenes en la Unimag. *El Heraldo (Barranquilla)*, 21 Nov.

ODH, (Observatorio de derechos humanos de la vicepresidencia de la República de Colombia). 2014. *Estadísticas de violaciones a derechos humanos y Derecho Internacional Humanitario (1990-2013).* Bogotá: Vicepresidencia de la República.

Peña-Huertas, R., L.E. Ruiz, M.M. Parada, S. Zuleta, and R. Álvarez. 2017. Legal Dispossession and Civil War in Colombia. *Journal of Agrarian Change*, 17(4), 759–69.

Policía Nacional de Colombia. 2008. Secuestro en orden cronológico según autor (1990-2007). *Revista Criminalidad*, 50(1), 263–8.

Quinche Roa, J.M., P. Perdomo Vaca, and J. Vargas Reina. 2018. *El despojo paramilitar en el Magdalena: el papel de las élites económicas y políticas.* Bogotá: Observatorio de Restitución y Regulación de Derechos de Propiedad Agraria.

Renán, W. 2007. *Contextualización del desplazamiento en el departamento del Magdalena y Santa Marta (1997-2007).* Santa Marta: Universidad del Magdalena, Vicerrectoría de Investigación, Fonciencias, Sección I, informe final convocatoria 2004.

Romero, M. 2003. *Paramilitares y autodefensas. 1982–2003.* Bogotá: Planeta.

Romero, M., ed. 2007. *Parapolítica. La ruta de la expansión paramilitar y los acuerdos políticos.* Bogotá: Corporación Nuevo Arco Iris.
Romero, M., ed. 2011. *La economía de los paramilitares.* Bogotá: Random House Mondatori.
Romero, M., Á. Olaya, and H. Pedraza. 2011. Privatización, paramilitares y políticos: el robo de los recursos de la salud en la costa Caribe. *In*: M. Romero, ed. *La economía de los paramilitares.* Bogotá: Random House Mondatori, pp. 15–74.
Salinas Abdala, Y., M.P. Hoyos, and A.M. Cristancho. 2018. *Tierra y carbón en la vorágine del Gran Magdalena.* Bogotá: Centro Nacional de Memoria Histórica.
Sarmiento, J.P. 2016. Justicia transicional sin transición: el caso de la masacre de Nueva Venecia. *Co-herencia*, 13(24), 181–211.
Valencia, L. 2009. Municipio y violencia paramilitar en Colombia 1984-2008. *In*: F. Velásquez, ed. *Las otras caras del poder. Territorio, conflicto y gestión pública en municipios colombianos.* Bogotá: Foro Nacional por Colombia - GTZ, pp. 121–78.
Verdad Abierta. 2013. La caída de un señor de las tierras. *Verdad Abierta*, 14 Mar.
Viloria de la Hoz, J. 2008. *Banano y revaluación en el departamento del Magdalena (1997-2007).* Cartagena: Banco de la República.
Volkov, V. 2002. *Violent Entrepreneurs. The Use of Force in the Making of Russian Capitalism.* Ithaca, NY: Cornell University Press.
Zamosc, L. 1986. *The Agrarian Question and the Peasant Movement in Colombia: Struggles of the National Peasant Association, 1967-1981.* Cambridge: Cambridge University Press.
Zelik, R. 2015. *Paramilitarismo. Violencia y Transformación Social, Política y Económica En Colombia.* Bogotá: Siglo del Hombre.
Zúñiga, P. 2007. Ilegalidad, control local y paramilitares en el Magdalena. *In*: *Parapolítica, La ruta de la expansión paramilitar y los acuerdos políticos.* Bogotá: Corporación Arco Iris, pp. 285–322.

3

AGRIBUSINESS ECONOMY AND EMBEDDED DISPOSSESSION

During the last decade, a great amount of scholarship documenting cases of violent dispossession in Colombia has been produced. However, most of this literature, including my own work (see, for instance, Grajales 2011, 2013, 2015), has been focused on instances of clear-cut dispossession; these were cases in which defenceless peasants had been expelled from land they had owned and farmed for decades. They could claim unequivocal ownership rights over their land, and most of the time, their communities had remained bonded through forced displacement and dispossession, something that allowed them to claim their right to restitution – in some cases even before the existence of the 2011 Victims Act.[1] Most of my colleagues in academia, as well as my contacts in NGOs and state agencies, have called these stories of dispossession 'emblematic cases'. I believe that the scholarly attention granted to them was a necessary and urgent matter when it began in the early 2010s. However, this book argues in favour of a more fine-grained understanding of land conflicts and dispossession, a comprehensive view that re-embeds violent land grabbing within the context of the trajectory of agrarian capitalism.[2] Such an approach, I will argue, must focus on grey areas, cases in which the antagonists are not necessarily paramilitary frontmen, and instead are often individuals who are able to proffer arguments that are legally grounded – if not morally defensible.[3] In this grey area, claims of dispossessed peasants over land that they consider theirs are not legally perfect. They might be blurry, as overlapping and competing rights tend to become enmeshed over time. They might be vague, with peasants never having held uncontested rights to the property. While no quantitative evidence can sustain my hypothesis, most land restitution claims in Colombia probably fall within this grey area. But the importance of paying greater attention to the grey area is not merely a quantitative concern. My view is that if the grey zone is characterized by ambiguous land rights, it is precisely because this very ambiguity is generated by the entanglement of legal and illegal – as well as coercive and non-coercive – modes of domination. Such a grey area is not an accidental consequence of the great complexity of property rights in a country such as Colombia (which, in this respect, resembles many other cases). Rather, it is a more direct consequence of the fact that outright dispossession is merely the most visible facet of a longer-term process, whereby an inequality regime is continually consolidated in the ordinary course of everyday affairs.

This chapter focuses on this grey zone through an analysis grounded in four case studies, all drawn from Magdalena's banana district, analysed from a micro-perspective. Taken together, these four cases tell a story of how land access and dispossession are determined by the forces of armed violence, but also by those of the market and the legal field, and offer a reflection on the way in which these factors are intricately interwoven.[4] In all four cases, the ability of the peasants to obtain access to land was highly dependent on economic cycles. Favourable circumstances for land redistribution were spurred on by the crisis that affected the agro-export economy between 1993 and 2000. The crisis also relieved some of the pressure on land and spurred activism from peasants and agrarian workers. Yet, the return of productive activities was accompanied by new cycles of dispossession, made possible by the availability of paramilitary violence entrepreneurs. Between 2000 and 2004, each of the four communities I will examine in this chapter was dispossessed of land that they considered to be theirs. The typical pattern included the murder of prominent members of the community and the intimidation and forced displacement of hundreds of people. In each case, the paramilitaries' objective was to empty the estates, enabling the start of agribusiness ventures within only a few months of the forced displacement. Overall, throughout this chapter, I will endeavour to analyse the embeddedness of land dispossession in the development and transformation of agrarian capitalism.[5] The first section of the chapter outlines the transformations in both economic activity and paramilitary groups' influence in the banana district. The second and third sections analyse two different sorts of ties between land, activism, and violence. The final section provides a discussion of the comparative findings and a conclusion to my overall argument.

Violence, booms, and busts in the banana district

In Magdalena, as in many other cases in Colombia and beyond, dynamics of violence are connected with cycles of economic boom and bust. This does not, however, mean that a one-way causal relationship can be established between agribusiness development and paramilitary violence. The production of violence by a multiplicity of actors follows very diverse patterns and drivers, and cannot be captured by such a facile and simplistic relationship. Instead, following the views of the work done by scholars such as Gill (2016) and Ballvé (2020), I will focus on how various actors (militia leaders, agribusiness investors, activists) struggle to connect or disconnect the political economy of production and the politics of paramilitary violence. Economic booms can create favourable contexts for collective mobilization. This, in turn, can provide fertile ground for the intervention of violence entrepreneurs willing to offer their services to repress an unruly labour force. At the same time, economic crises can have dreadful consequences on working people, but they can also create new spaces for collective action, as the power of capital owners is curtailed and the pressure on the land market is relieved.

Magdalena illustrates the complexities of these connections in various manners. From the end of the 1980s, the banana economy re-emerged after decades of gloom. The employment of thousands of banana workers called both for social mobilization – spearheaded by labour unions – and for the policing and repression of collective action, undertaken by local militias such as the Rojas and Giraldo groups in the name of anti-communism. Violence in the banana district reflected a larger pattern of murders in Magdalena, as Figure 3.1 illustrates. Homicides committed in the territory of the banana district consistently accounted for 40 per cent of the department's total violent deaths in the early 1990s.[6] The economic boom was short-lived. In 1993 and 1994, a combination of climatic conditions and global trade relations caused a violent downturn in the banana business. Between 1994 and 1996, banana production in Magdalena fell by 40 per cent. A decrease in the levels of violence is also observable during those years. Murders committed in the banana district accounted for only 29 per cent of Magdalena's total. While at the scale of the banana district, this might be linked to a lesser need for labour repression (as many plantations were shut down), the tendency corresponds to a regional pattern, which suggests other important reasons, such as changes in the political economy of drug trafficking. However, the effects of the crisis were also felt in other ways. The meltdown drove businesses into bankruptcy and caused huge losses for some of the province's great fortunes, while plunging small peasants into extreme hardship. Nevertheless, it also relieved some of the pressure on land and spurred

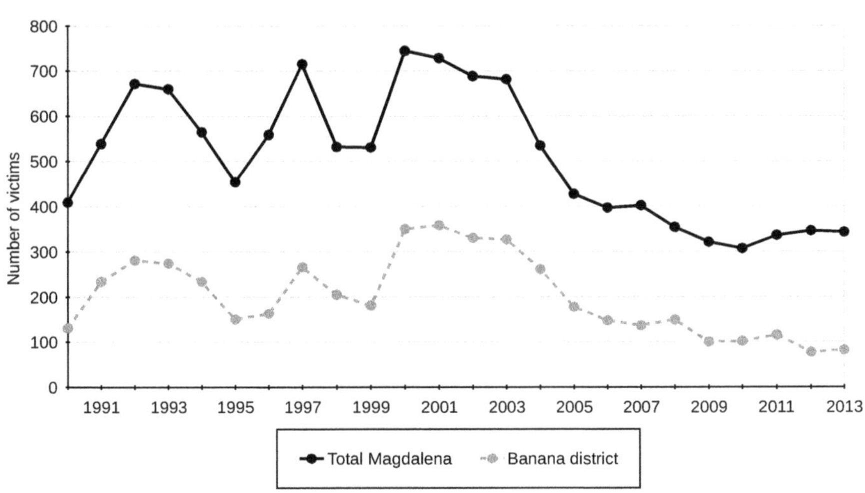

FIGURE 3.1 Murders in Magdalena and the banana district. Source: the author, data from ODH (2014)

activism from peasants and agrarian workers. The years of the crisis corresponded to a new wave of activism, as both landless peasants and laid-off farm workers fought for the right to farm their own land.

Two new waves of violence were felt in the following years, corresponding to the arrival of the AUC's militias in Magdalena. Murders peaked in 1997 and again in 2000, with each surge reflecting an additional phase of the AUC's expansion. In the banana district, violence was particularly brutal in the early 2000s. Most of the peasant collectives that had been formed in the years of the crisis for the purpose of reclaiming land were repressed by the paramilitaries. This corresponded to the return of economic growth, associated with the recovery of the banana economy and the palm oil boom, in addition to a much more aggressive strategy of territorial control by the militias not only in the northern lowlands, but in all of Magdalena.

This section will successively present the transformations in paramilitary violence in the banana district and the variations in the economic context. The remainder of the chapter will use a series of case studies to connect these two factors while reflecting on the economic embeddedness of the militias.

The social order of the plantation

The banana district, Colombia's top global agribusiness hotspot, had fallen into decay in the late 1960s, when the United Fruit Company moved its operations to the region of Urabá, some 500 kilometres south-west along the coast (Bucheli 2005). In the 1970s and 1980s, Urabá became a paradigmatic enclave, with thousands of workers living in very poor urban ghettos with inadequate sanitary conditions, and most profits being captured by the alliance between a transnational corporation and the domestic elites of Medellín who controlled most of the land (Ortíz Sarmiento 2007; Ballvé 2020). This offered a particularly favourable environment for guerrilla activism, and both the FARC and the Maoist EPL saw Urabá as a key battleground for political struggle. Violence escalated quickly, as plantation owners and local military units resorted to paramilitary militias to kill, torture, and harass anyone suspected of harbouring guerrilla sympathies (Botero 1990; García 1996; Suárez 2007). However, this did not diminish the left's political influence, as the UP would become the top political force in the region by the late 1980s (Carroll 2011). By then, Urabá was characterized both by a fast-changing political landscape and by egregiously high indicators of political violence. Both were bad for business.

As a consequence of the social and political turmoil in Urabá, various foreign corporations and a handful of Colombian investors began moving some of their operations to Magdalena in 1988. By 1990, banana production in the Santa Marta area had quadrupled, reaching more than 270,000 tons (Bonet 2000). In 1991, the sector employed more than 9000 people. A single village in the heart of the banana district, Orihueca, was home to over 3000 commercial farm workers (El Tiempo 1991a).

With the banana boom came a surge in paramilitary violence throughout the banana district. Between 1989 and 1995, 74 per cent of the murders committed by paramilitary groups in Magdalena took place in the agribusiness zone. In 1992 alone, the militias murdered 54 people, many of them plantation workers.[7] On 22 July 1991, a group of paramilitary troops claiming to be members of the Rojas group murdered three banana workers, two men and one woman. Throughout the year, the paramilitaries had been circulating vocal threats against unionists, accusing them of being 'guerrillas disguised as civilians' (*guerrilleros en civil*). The climate of fear pushed the Sintrainagro workers' union (*Sindicato nacional de trabajadores de la industria agropecuaria*, National union of agro-industry workers) to organize a national forum in November 1991 aiming to raising awareness about paramilitary activity in the banana district. A week after the meeting, the four members of the local chapter who had travelled to the forum were murdered. In 2009, during a court hearing, the Rojas brothers acknowledged having committed these crimes, and attempted to justify them by arguing that the victims were guerrilla supporters.[8] Most of these murders resulted from the will to tame, or even eliminate, trade union activity in the region. As a matter of fact, with the rise of the banana economy, Sintrainagro was embarking on a massive recruitment campaign. According to a union leader at the time:

> We lacked many basic supplies, we lived in very poor sanitary conditions, and we worked without counting our hours! As a result, it was a favourable environment for people to organize and fight for their rights.[9]

However, workers' unions aroused mistrust and rejection. As another labour unionist living in Cienaga, says: 'They [the entrepreneurs] think that we unionists are like a weed that grows in their company and push it to bankruptcy. That's why, when a worker joins a union, he gets fired from the company'.[10] Veiled threats against union activists were common, such as in this quote attributed by the Constitutional Court to one of the richest banana entrepreneurs in the area:

> I don't like unions [...] The other day a bunch of guys from the self-defence [paramilitary] groups came to my office and they asked me for the list of all the people who are members of the union, the leaders in particular. Well, you see, they don't like unions either.[11]

The consolidation of paramilitary troops' grip over the entire Magdalena region was violently felt in the banana district. ACCU and AUC militias began to operate in the vicinity of the city of Ciénaga in 1996. They partly relied on existing gangs to expand their influence. Much as it had been the case for their predecessors, the violence doled out by the AUC was massively aimed at breaking up workers' unions – particularly Sintrainagro, which they saw as a key political enemy. This new wave of assassinations can be said to have begun in July 1997, when AUC militia members murdered two Sintrainagro union leaders, Camilo Suárez and Mauricio Tapias. In the second half of the year, several union activists, including Sintrainagro's vice-president Amaulis Mejía, were also murdered. By the end of 1997, 21 banana workers had been killed by the AUC (CINEP 2004). In August, a Red Cross representative sounded the alarm, considering that northern Magdalena was becoming a hot zone for paramilitary militias (El Tiempo 1997).

The paramilitary grip on the plantations and villages of the banana district grew even stronger in 2001, when all the region's paramilitary groups were brought under the exclusive leadership of Jorge Cuarenta. In January 2001, José Luis Güette, Sintrainagro's president, was murdered. By 2001, Sintrainagro, a union that had reached 2600 members in 1994, was reduced to about 900 members in all. The violence was not only aimed at the union's leadership, but was intended to terrorize all banana workers to dissuade them from any attempts at collective action. One of the individuals I spoke with interlocutors recalled the following:

> There's a place called 'La vuelta del cura' (the priest's turn) near Sevilla. That's where they used to leave the bodies, because the bus line that transported the workers had to pass there. Every morning when we went to work, we discovered dead people on the way.[12]

What motivated these egregious acts of violence? According to the area's paramilitary commander, José Gregorio Mangones Lugo, aka 'Carlos Tijeras', the brutal campaign against unionism was a response to the threat that workers' organizations – and particularly Sintrainagro – posed to the paramilitary groups' main clients: banana industrialists. In court testimony, he has consistently depicted the relationship between paramilitary groups and banana corporations as one of providing security, in the broad sense of the term:

> After we established a quiet environment and became the local law enforcement agents, Chiquita and Dole plantation managers depended on us to respond to their complaints. My men were regularly called upon by Chiquita and Dole plantation managers to provide help in the face of criminal activities or to fix other problems […] In most cases, those who were eliminated were unionists or people intending to claim land ownership [over land covered by the commercial farms].[13]

Do other sources confirm the nature of this relationship? As early as 1990, local observers acknowledged a link between the arrival of entrepreneurs from Urabá, where violence in the plantations had been escalating for years, and the surge of paramilitary activity in Magdalena. Interviewed by a national newspaper, the chief of police operations in the department considered that 'entrepreneurs arriving from Antioquia want to introduce the paramilitaries to neutralize the unions' (El Tiempo 1991b). Similarly, a 1990 policy white paper claimed that: 'the new capital […] brings with it a pattern of relations between workers and bosses based on repression and the use of paramilitary forces' (CORPES 1990). The close ties between paramilitary groups and banana entrepreneurs were not unknown to state authorities. A military intelligence report from 1998 claimed that every one of the banana district's entrepreneurs was financially supporting the Rojas. Some of their clients allegedly ordered the murder of a local Sintrainagro leader and five activists in February 1994.[14] The Rojas were not the only paramilitary group active in the zone. The Durán family gang, a group of violence entrepreneurs from Fundación, frequently came up in the investigations into murders of labour unionists (Renán 2007). In other cases, union leaders and activists were threatened, harassed, and murdered by criminal gangs who, although they were primarily active in other areas of criminal activity such as theft and racketeering, were willing to sell their services to bosses facing labour conflicts. Gangs such as the Meriños and the Cuquecos, primarily active in the town of Ciénaga, fell into this latter category (Martínez and Renán 2011, 396).

The testimony of paramilitary commander Carlos Tijeras has been corroborated by internal documents from Chiquita Brands. In 2007, the firm entered into a plea agreement with the US Federal Department of Justice (DOJ) in relation to criminal proceedings against top company executives for their part in funding the AUC. The proceedings exposed them to significant criminal liability, in particular because the militia had been labelled as a foreign terrorist organization in September 2001. In compliance with the plea deal, Chiquita turned over thousands of documents to the DOJ, which traced payments to the guerrillas and the AUC in relation to the company's operations in Urabá and Magdalena. These documents were made public through a Freedom of Information Act request made by the Washington-based National Security Archive. They were published in 2011 by the Archive, and provided evidence for payments to the AUC amounting to 1.7 million dollars (Evans 2011). While Chiquita and other corporations characterize the payments as 'extortion', several NGOs, victims' organizations, and law clinics both in Colombia and the US have filed lawsuits against Chiquita executives. The plaintiffs allege that Chiquita was no victim, as it benefitted from the paramilitaries' services in policing the plantations, taming protests, and securing access to disputed land. The bulk of Chiquita's internal documents supports this thesis. For instance, in one internal memo on the company's Magdalena operations, it was stated that the corporation 'should continue making the payments', as it 'can't get the same level of support from the military'.[15] At the time of my writing, these judicial proceedings are still pending. Similar complaints have been filed against Dole Food Company, but much less information has been made public on this firm's activities, and the company has repeatedly denied any link to the paramilitaries.

AUC's arrival in the banana district represented a change in the scale of these relations. Beyond the specific case of multinational corporations, both the 'Chiquita papers' and Tijeras's testimony depict a standardized financing network with money flowing from banana export companies to AUC commanders' coffers. The militia charged a fee corresponding to USD 0.03 for each banana box leaving the plantations. According to Colombian police investigations, this represented around 70,000 dollars per year for the paramilitary unit in charge of collecting the fees in Northern Magdalena.[16] The money was paid out to a front-office firm, which acted as an intermediary between the corporation and the paramilitary group. Questioned about the services provided in exchange for that money, Tijeras testified that:

> We got rid of the guerrillas who were extorting them, kidnapping their managers, and burning down their plantations, and we offered protection from the criminal gangs who stole their supplies and equipment, robbed the fruit shipments, and looted the plantations [...] We guaranteed the return of invaded plantations and the stability and the real-estate value of the plantations, we did political-military work to expel guerrillas from unions and social organizations, and we escorted the trucks transporting the fruit from the plantations to Santa Marta's harbour.[17]

As the above citation illustrates, recovering 'invaded plantations' was one of the potential services that paramilitary commanders saw as falling within their purview. Later on in the chapter, I will focus on stories of such 'invasions'. In reality, these were disputed estates, where peasants were asserting a right of access to land also claimed by large landlords. The presence of paramilitary violence entrepreneurs such as Tijeras provided the means for repressing peasant claims. Before moving on to this specific point, I will present an overview of the economic context during those years, which dictated the possibilities for collective action and the demand for violent services.

Global banana wars and the local economy

These transformations in the paramilitary rule over the banana district took place against the backdrop of a changing economic context. The 1990s banana crisis was caused by a combination of economic, climatic, and biological factors. It lasted several years, in spite of international trade negotiations and financial support from the government to local producers. In the aftermath of this first downturn, the contagion of the South-Asian crisis sent the entire Colombian economy into a freefall beginning in 1998. Agribusiness activity in Magdalena only resumed in 2000, fuelled by both bananas and palm oil. This subsection briefly sketches how the crisis unfolded.

The crisis' primary cause was the limited access of Colombian producers to the European market. Historically, bananas have been a sensitive crop for members of the European Community. This is the result of opposition between member states' competing interests. A protectionist bloc was formed by countries such as Spain, Portugal, and Greece, which all have domestic production that they sought to protect from competition from cheaper Latin American imports. They joined forces with former colonial powers such as the UK and France (which also has a domestic production in the Antilles), as well as Italy, which has a stake in the Somalian banana industry. These governments had diplomatic interests in offering preferential access to their markets for former colonies, which formed the so-called ACP (Asia, Caribbean, Pacific) group. ACP production was vulnerable to international competition, as the group's productivity was generally lower than that of Latin American producers. In the opposite camp, a free-market bloc was gathered around Germany, Austria, and the Scandinavian countries, with Benelux nations somewhere in the middle (Tangermann 2003a). These liberal economies represented a huge market for Latin American producers. However, market access was dependent on differential trade regimes, which stood in open contradiction to the EU's unified policy.

National trade regimes were safeguarded until the early 1990s, when it became clear that they were incompatible both with the evolution of the common market and with global trade regulations. However, the European Commission faced multiple constraints in refashioning a banana trade regime: common ground had to be reached between protecting domestic production, honouring past commitments with ACP countries, and dealing with the necessity to produce a minimally acceptable outcome for all the other interested parties (Tangermann 2003b). This was made even more complex by the fact that these interested parties included not just Global South governments and producers, but also two massive US banana firms, Chiquita and Dole, who controlled most of the export and distribution of Latin American bananas. These two companies had significant political influence in Washington and were committed to using it to safeguard their access to the European market (Bucheli 2005).

In December 1992 and January 1993, the European Commission established a new trade regime. It was based on a tariff rate quota (TRQ), composed of two distinct quotas: a duty-free quota for ACP-country banana producers, and a quota for non-ACP countries, which applied to Latin American producers. For this second group, imports within the quota were charged a tariff of 100 ECUs[18] per tonne. Imports above the quota were charged a prohibitive tariff of 850 ECUs per tonne (Tangermann 2003a, 37). The trade regime was transformed over the following years, as a more open market was made possible by harsh US opposition to the tariffs, lobbying from Latin American banana producers at the WTO (World Trade Organization), and internal struggles in the EU. These 'banana wars' encapsulate many of the inherent contradictions of the contemporary system of international trade, underscoring the conflicts between the interests of multinational corporations, domestic production, and a diversity of developing countries (Josling and Taylor 2003).

For Colombia, the banana wars resulted in an initial shock in 1993, as many producers were greatly dependent on access to German and Scandinavian markets. Shortly after the shock, the Colombian government initiated attempts to negotiate better market access conditions with the EU. The administration's joint efforts with five other Latin American countries resulted in a more favourable agreement signed in 1994 (Tangermann 2003b). However, Washington opposed this accord, as the Clinton administration considered it to be detrimental to the interests of US corporations. As the Commission itself was having a hard time obtaining the support of the European Council and the Parliament, the years that followed were characterized by constant instability in the rules applicable to the banana trade. This

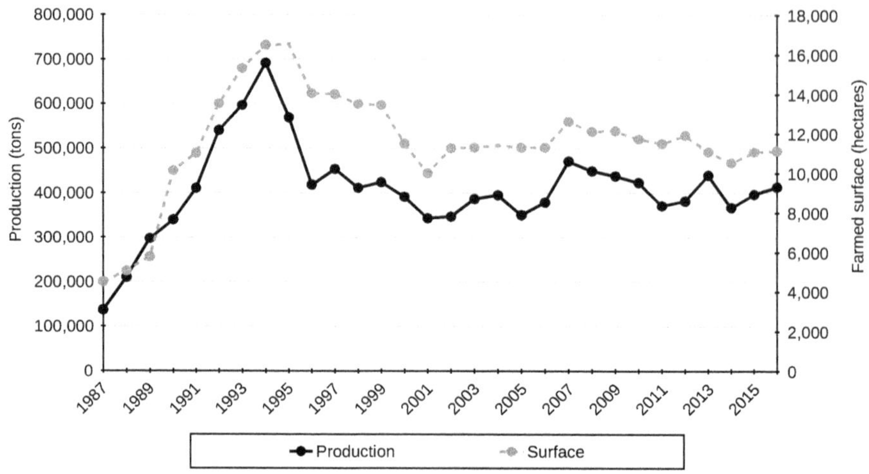

FIGURE 3.2 Banana production and farmed surface. Source: the author, data from MADR (2020)

was a challenge for the Colombian government, whose efforts to support producers were repeatedly undermined by the changing regulatory framework (Bonet 2000).

In Magdalena, the effects of the commercial backlash were magnified by unexpected and excruciating sanitary and weather conditions. In 1993 and 1994, an outburst of black sigatoka, a fungus that attacks the leaves of banana trees and one of the most common threats to this crop, affected the plantations of Magdalena. In August 1993, tropical storm Bret struck the coast of Colombia and destroyed hundreds of hectares of banana plantations in northern Magdalena. The years that followed were characterized by a succession of extreme climate phenomena such as the 1996 wet season and its violent rains and storms.

In response to this multifold crisis, the Colombian government implemented various strategies to support the banana industry. A credit line was opened through state-owned credit provider Fidufi, which benefitted 70 per cent of Magdalena's banana producers (Viloria de la Hoz 2008, 40–1). An export subsidy was put in place in the form of a tax cut. Special credits for infrastructure and sanitary investment were also made available (Lombana 2012, 10). However, in the final years of the decade, these efforts were made vain due to the violent meltdown of the country's economy. In 1999, largely as a consequence of the Asian financial crisis, the Colombian economy went into recession, with a negative growth rate of 4 per cent – the worst crisis in the 20th century (García-Echeverría 2019, 282). The estimates at the subnational level show an even worst situation in Magdalena, as the economy contracted by 5 per cent that same year (Romero-Prieto 2007, 11).

Agro-business activity began to resume in 2000 and 2001. By 2002, production was again on the rise. But bananas were no longer the dominant crop. Throughout the banana district, competition from palm oil production was pervasive. In 2002, there were 52,986 hectares of oil palm plantations in the entire area, compared to a

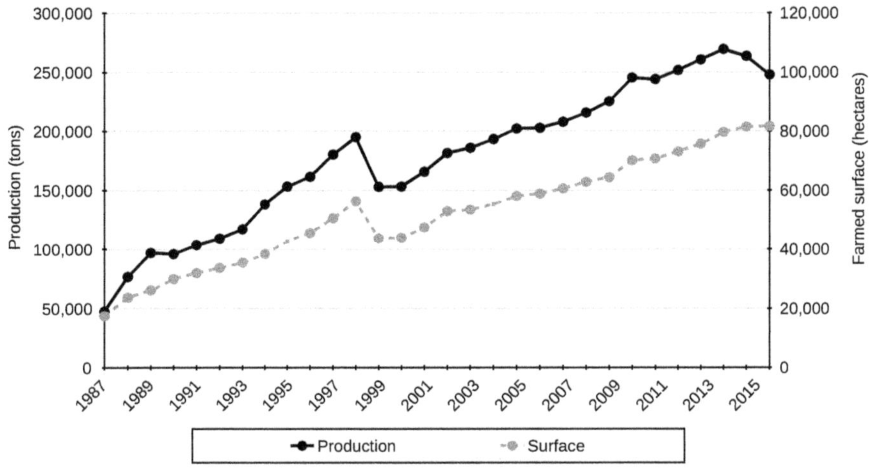

FIGURE 3.3 Palm oil production and farmed surface. Source: the author, data from MADR (2020)

mere 11,267 hectares of banana plantations. While the former is a crop that requires much more room than that latter, Figures 3.2 and 3.3 show that banana production never resumed at the levels of the early 1990s boom, while palm oil continued its progressive rise. However, the stability of the banana business conceals a phenomenon that is difficult to quantify: the concentration of banana farms. Today, the largest 10 per cent of estates (farms over 55.8 hectares) concentrate 78.4 per cent of the land in the municipality of Zona Bananera;[19] the top 1 per cent (14 estates) alone accounts for 45 per cent of the land (DANE 2016). Interviewed agro-export professionals claim that after the 1990s crisis, the sector became much more concentrated, with a few firms holding most of the land. To the extent of my knowledge, this claim is impossible to corroborate: the most reliable source is the 2014 agricultural census, but it cannot be compared over time as the previous one dates from the 1970s, cadastral registries in the area are out of date, and large landlords usually own a conglomerate of firms, each of them being the legal owner of only one or two estates. The intricate legal schemes used to conceal the names of the actual owners make it particularly challenging to trace the transformations in land ownership. However, indirect indications support the concentration hypothesis. For instance, as the book's closing chapter will show, there was a sudden surge in investment in private irrigation facilities after the banana crisis, something that can only be supported by high financial capacities.

Organized labour and global value chains

When the mid-1990s banana crisis took root, numerous plantation owners went bankrupt and dismissed their workers. In some cases, this resulted in new opportunities for accessing land. This section analyses two cases in which workers at failed plantations sought to gain control over their means of production and become independent farmers. While they managed to thrive in a very troubled economic environment, they were ultimately dispossessed by landlords claiming to have legitimate ownership rights. In both cases, dispossession required a combination of force – provided by paramilitary militias – and legal manoeuvring. Both cases show how fragile the workers' claims were, and how the protection of property rights worked in favour of powerful actors.

La Marcela

La Marcela is a 210-hectare estate located near the village of Varela, a part of the municipality of Zona Bananera. In the late 1980s and early 1990s, it belonged to Armando Dangond Noguera, a powerful member of the local elite and a thriving banana entrepreneur. La Marcela was one of the most modern estates in the area, equipped with state-of-the-art facilities and employing up to 170 workers during the harvest season. As a matter of fact, the Dangond family played a critical role in the early 1990s banana boom. The family firm, Arcemar Limited, was created in 1988 and invested massively in the banana business during those years. In addition to their economic relevance, the Dangonds were politically influential. Two of the Dangond brothers were active in politics: Orlando had been elected to the department assembly in 1984 and Victor won a seat in the House of Representatives in 1986. However, the Dangonds' reputation was far from pristine, as in 1988, Victor had faced criminal charges for the murder of two Sintrainagro leaders killed by paramilitary hit men. Victor Dangond was tried but he was ultimately acquitted.

The banana crisis put an end to the estate's prosperity. La Marcela's workers were covered by a collective labour agreement that expired in 1994. In October 1993, the estate's workers adopted a package of demands for the upcoming round of bargaining. This included recognizing Sintrainagro as the representative of the entire labour force and the delivery of permanent contracts to day labourers regularly working at the estate. However, the collective bargaining never took place, as in April 1994, the firm owning the estate filed for bankruptcy and stopped paying workers' salaries. The workers declared a strike and blocked the harvesting activities. The estate's owners attempted to send a contractor commissioned to harvest the existing bananas, but the workers barred entry to the plantations.

The workers' collective efforts then moved to the courts. They were supported by Sintrainagro, which provided a lawyer and accompanied the workers' spokespeople to the labour inspection hearings. A lawsuit was filed in the employment tribunal in March 1995 in which the workers asked for the payment of back wages and the settlement of their employment contracts. The court ruled in favour of the plaintiffs between October 1996 and May 1997. However, during the course of the proceedings, the workers strived to find a way to provide for themselves. With the support of Sintrainagro, they received authorization to keep the existing harvest for themselves, as it was overdue and was no longer apt for export, and used it both for their own consumption and to make a profit by selling it at the local markets. Nevertheless, this only provided temporary relief. In addition, the workers were worried about other landless peasants settling inside the estate, as many saw the company's bankruptcy as an opportunity to occupy small plots of land left idle by the crisis.

In response to the abandonment of the estate by its legal owners, the workers felt entitled to farm the land. They created three distinct groups. The first group was formed by 34 people (only one woman) and included most of the union activists who had been at the forefront of the political and legal struggle. They called themselves the group of the 30 and decided to jointly farm an area of about 80 hectares with the intention of producing and selling their own bananas. The second group, primarily made up of former administrative employees from the banana company, opted for individual farming of plots no bigger than 4 hectares. They occupied a total area of about 100 hectares. A third group, composed of family members of the workers and much more feminine in its composition, jointly farmed the edges of the banana plantation, and were dedicated to the production of food crops (cassava, maize, plantain, pumpkins, etc.) that were locally marketed. Most of the farmers did not settle inside the estate, as they lived in the town of Varela and commuted every day to La Marcela.

The group of the 30 created a cooperative in August 1995 with the support of Sintrainagro. Despite multiple difficulties stemming from the banana crisis and the bad weather conditions of 1995 and 1996, the cooperative managed to thrive. It benefitted from credit provided by a local producers' association and obtained access to trade circuits to export its produce. The cooperative not only distributed income to its members, but it also generated jobs, as additional labourers were needed during the harvest season. As one of the members of the group of the 30 remembers:

> And so we were moving on. We even had employees, they came for a day and a half of harvesting, we began to have a positive influence in the economy of [the village of] Varela. On payday at [the cooperative], you could see the movement, the money, the joy.[20]

In 1998, a labour judge ordered the sequestration of all the Dangond family's company assets, a procedure that prevented the original owners from selling the estate without duly compensating the workers. By the year 2000, both the cooperative and the independent farmers were prospering. They considered La Marcela's land as their own, and were hoping to obtain the legal recognition of their claim. The legal proceedings were not yet finished, but the courts had systematically ruled against the former owners.

Diana María

A similar process of labour and land struggle, coupled with the capacity of a collective of agrarian workers to organize and take possession of their bosses' land, can be observed in the case of Diana María, a banana-producing estate located near the village of Santa Rita in the municipality of Zona Bananera. The estate belonged to the Fernández de Castro, a powerful family of landowners from Santa Marta. They had purchased it from the United Fruit Company's subsidiary in 1961, when the multinational corporation had relocated its activity from Magdalena to Urabá. Diana María prospered in the late 1980s banana boom. By then, it belonged to the daughter of its original owner, Teresita Fernández de Castro. This was a family of banana entrepreneurs, as Teresita's husband, Carlos Manuel Dangond, belonged to the powerful Dangond family who owned La Marcela and many other thriving plantations of the region.

The crisis of Diana María combines elements from the local economy with the personal history of the estate's owners. In 1991, Carlos Manuel Dangond was killed in a car accident. Probably due to this family loss, the management of the estate deteriorated. There were delays in the payments of salaries and neglect in the maintenance of crops and machinery. This drove Teresita Fernández to rent out the estate to the Colombian corporation Banacol in 1993. Shortly afterwards, the contract was terminated due to conflicts between the two parties. When the banana crisis struck the region, Fernández de Castro claimed to be unable to pay the workers back wages or to resume the administration of the estate.

From then on, the story of Diana María followed a very similar path to that of La Marcela. Sintrainagro helped the workers file suit against their former employer. This pushed Fernández de Castro to authorize them to farm seasonal crops that would provide them a momentary source of food and a basic income, but he forbade them to plant bananas. The workers occupied the estate in 1994 and 1995, collectively producing food crops such as beans, cassava, and maize and holding up their end of the bargain. In 1998, a labour judge ruled in favour of the workers, ordering Fernández de Castro to pay them back wages in the amount of COP 554 million (410,000 US dollars at the time). An appeals court ruling confirmed this judicial order. The businesswoman reacted by claiming that she did not have the money to pay the arrears and that the workers should keep the land as payment. However, she never formalized this transaction, as she told the workers' representatives that her own family would not let her abandon her father's land. This is the point where the Diana María workers decided to follow a similar strategy to the one used by La Marcela workers. One of these workers-turned-peasants remembers:

> That's when we decided to stop watching over the estate and start cultivating. We met and we told the people that the lady had no money, and that we were not going to leave the estate. The people accepted and we agreed to go and farm. The next day each one of us measured a piece of land… We did not draw up a document; we just agreed and divided up the land. We didn't fence off the plots.[21]

A cooperative was officially created in 1999 with 22 members. While it was initially meant to be a vehicle for farming in Diana María, it also integrated other small producers from the surrounding area. The cooperative was not as well-organized and prosperous as La Marcela's, but most of the farmers managed to live off their land, selling bananas and growing their own food.

Curbing collective action and vacating land

Between 2000 and 2001, the people of La Marcela and Diana María were dispossessed in a dreadful series of events including murders and threats by paramilitary militias and legal manoeuvring to dispute their legal claims. The new landlords endeavoured to erase any trace of the past struggles, instead labelling the workers and peasants as squatters who had taken advantage of the economic crisis to invade the plantation.

By 2000, La Marcela workers had won every legal battle they had engaged in with their former bosses; in spite of a nationwide recession, the cooperative's business was thriving, and the community was probably on the brink of obtaining legal ownership over the land they had been farming. However, in a matter of just a few months, the entire community was expelled from the estate and new owners took possession of the land with the support of paramilitary troops.

In March 2000, a group of people presenting themselves as representatives of the new owners of the estate told them that their former boss had sold the land and that they were required to leave. The workers refused to comply and reported the visitors' threatening behaviour to Santa Marta's attorney general's office. On 8 September, a group of 30 men belonging to the AUC arrived at the estate in the early morning. They held all the workers who were present and killed four of them who were among the leaders of the cooperative. The community was told to vacate the estate. Two months later, some of the peasants who had settled in the neighbouring villages were told that production was set to resume and that they could get a job with the new owner. Some accepted. A few days later, yet another massacre claimed five additional lives in the neighbouring village of Varela. The workers who had returned to the plantation attracted by the job offers ultimately fled out of fear for their lives.

The Diana María community went through a similar series of events. By 2000, the peasant collective was producing export-quality bananas and selling them through a small producers' cooperative. In 2001, the brother of the original owner contacted the community. He urged them to leave the estate, declaring that he considered them to be squatters and offering to take the existing machinery as a payment for their debts. When they refused to leave, he threatened them, telling them to be careful because 'security was quite bad in the area'.[22] The peasants sought the protection of the People's Ombudsperson (*Defensoría del Pueblo*), a public office tasked with the protection of fundamental rights. However, Magdalena's Ombudswoman at the time was later questioned in relation to her supposed partiality, as she would go on to pursue a political career sponsored by a former senator who had been convicted for his links with paramilitary militias (El Heraldo 2011).

On 12 September 2001, four paramilitary soldiers entered the estate and killed two members of the community. The occupants were told to leave within 72 hours. Some found refuge in neighbouring villages, but many members of the community decided to take their chances in the cities of Santa Marta or Barranquilla, or even to make the journey to Venezuela, a prosperous economy at the time. A month after the occupants of the Diana María estate were expelled, the existing crops were uprooted, the estate was fenced in, and new machinery was brought in to produce palm oil. As of today, there are 97 hectares of planted oil palm trees in full production. In 2014, the paramilitary commander of the area, Carlos Tijeras, declared that he ordered the killing of Diana Maria's workers because the estate's owner came to one of his lieutenants accusing the people on the farm of being guerrilla sympathizers and land invaders.[23]

In both cases, further legal manoeuvring was necessary to legitimize the dispossession from a legal standpoint. The two primary obstacles to free exploitation of the estates were the sequestration order that had been issued by labour judges (as security for the payment of back wages) and the pending litigation, which might conceivably result in the expropriation of the two commercial farms in favour of the workers. From the outset of the forced displacement, various steps were taken to eliminate these two obstacles. Legal counsel for both communities was provided by the same lawyer, who regularly worked for Sintrainagro. He seems to have been instrumental in ensuring that the two estates were lawfully seized from the communities occupying them.

In the case of La Marcela, the manoeuvres began a few months after the September 2000 massacre. In June 2001, the workers' lawyer addressed a filing to the judge in charge of the case. He withdrew the lawsuits for the payment of back wages and asked for the sequestration order to be lifted. This request was complied with in a matter of days. In July 2001, the sale of the estate could be legally recorded. Between that time and late 2007, the estate changed hands five times. Since then, the plantation has belonged to one of the largest banana producers in the area, which changed the name of the farm to La Quinta in December 2015. These numerous transactions make it particularly difficult to establish the responsibility of the current owner.

Diana María features a similar series of events. The forced displacement of the community happened in September 2001. In February 2002, the estate's new owner allegedly ordered all the displaced peasants to convene at a notary's office in Santa Marta, where they were told to sign some papers. The documents were powers of attorney and payment receipts. In exchange, they were given small sums of money, between 50 and 250 US dollars each. They now claim to have signed the documents out of fear, suspecting that the new owner had something to do with the murder of their two comrades. The documents were later used by their own lawyer to withdraw the lawsuits against their former employer in the exact same manner as in the La Marcela case.

The intertwinement of violence and legal manoeuvring in these two cases reflects a fundamental aspect of the struggles on which the workers of the two estates embarked on in the middle of the 1990s. If they managed to turn an unfavourable economic context into an opportunity for access to land, it was only made possible by a tremendously well-organized mobilization that enjoyed the support of the workers union and by turning the law in their favour. However, their success made them dependent upon an extraordinary conjuncture. While the wave of paramilitary violence of the late 1990s and early 2000s might have seemed like a radical rupture in the social order, in many respects it amounted to a return to normality. Mobilizations such as those of La Marcela and Diana María certainly posed something of a threat to the existing social order – a challenge to the inequality regime. The actions of the paramilitary groups in these cases did nothing but reinstate the imperilled status quo. The next section will pursue this demonstration based on two further cases of land dispossession.

Agrarian reform in times of economic crisis

In a second pattern associating economic cycles and land access, the relationship is indirect. This concerns cases in which the slump in the banana economy created a favourable context for the allocation of land through the agrarian reform policy. By the mid-1990s, the most important route to land allocation was a market-led mechanism (see Chapter 1) intended to make it possible for landless peasants to purchase land. The policy consisted of a subsidy covering 70 per cent of the land acquisition cost, with the remaining 30 per cent being paid by a loan from the Agrarian Bank (*Caja Agraria*), a state-owned banking corporation. Incora (the Colombian institute for agrarian reform) acted as an intermediary for the land purchase. An alternative mechanism, much less common and quite a reminiscence from the peasant struggles of the 1930s and the 1960s–70s was the expropriation of idle land. From a legal standpoint, the peaceful occupation of idle land for more than five years provides a ground for claiming property rights.

In both cases, land access is highly dependent on economic cycles. In the first case – market-led land access – potential land sellers are more willing to sell-off land when commercial agriculture yields are poor. As a significant share of the transaction is covered by a subsidy, a seller could expect to get a better price on the assisted land market than on the open market (Deininger 1999; Machado 1999). In the second case – expropriation of idle land – the very fact that land can remain idle in agribusiness regions like northern Magdalena is spurred by economic crises.

Moreover, these two routes to land access posed different challenges to their potential beneficiaries. In the first case, a lack of technical and commercial support might turn the loan used to purchase the property into an unbearable debt burden (Machado 1999). In the second case, the expropriation of a commercial farm – even if the land was unexploited and the owner was absent – required colossal efforts and was highly uncertain (Londoño Botero 2016). Decades of political debates on the agrarian issue had resulted in overarching safeguards provided to landowners against expropriation, as shown in Chapter 1. Moreover, local Incora offices were seldom eager to engage such long and excruciating procedures, particularly when the claim was being made by impoverished peasants and the party risking expropriation was an influential landowner. The two following case studies illustrate how these already precarious routes to land access were made even more hazardous by the intervention of paramilitary militias.

Tierra Grata

In 2018, I drove to Guacamayal, a dusty village in the heart of the banana district, to meet with the current leader of the community of Tierra Grata. The story of this peasant collective epitomizes the challenges faced by countless peasants in their attempts to make the assisted-market policy work in their favour. In 1996, after a series of mobilizations, a peasant association was formed under the leadership of Samuel Cervantes, a man who had been instructed in collective mobilization through his membership in ANUC (the National Peasant Association). A group of 20 families managed to obtain Incora's support for the purchase of two estates totalling 500 hectares in the municipality of Pueblo Viejo. The land was primarily dedicated to cattle breeding and it belonged to two large owners in the region. The process of adjudication is symptomatic of the shortcomings of land allocation through Incora, and its consequences weighed heavily on the peasant community's ability to retain control over their land afterwards. As one of the leaders of the community remembers:

> Some of the agrarian reform beneficiaries had been there from the beginning of our struggle, or in a few cases were family members of the association's affiliates. But others were put on the title deed in exchange for their help with the procedure. For instance, a guy from the land registry office of Ciénaga ended up as one of the beneficiaries. We had to include him or else we wouldn't have been able to move forward with the paperwork.[24]

At first glance, this account seems only to reveal the corruption that is rife within the local government offices in charge of land management. More importantly, however, it tells of the difficulties faced by peasants, who, despite complying with the law in every way, must still seek out other resources to effectively assert their rights.

The estate was jointly allocated to the members of the association in September and October 1996. This did not imply a lack of individual rights. Shortly after the title was signed, the land was divided among them. The association allocated plots of approximately 13 hectares for each family. Part of the estate was left undivided and reserved for cattle grazing. Two community halls were built and several families constructed their homes and wells on their plots. Farming commenced with subsistence crops such as pumpkin and maize. In accordance with the agrarian reform policy of the time, the people of Tierra Grata were entitled to a line of credit intended to support their agricultural activities. Initially, they had planned to produce palm oil, as such plantations were soaring at the time in northern Magdalena and collection and extraction facilities existed near the estate. However, for reasons that my research did not reveal, the loan was ultimately used to purchase cattle. Each family received 12 heads of cattle. However, the community lacked any experience in cattle breeding, and no state agency provided technical or sanitary support. The variety of the animals was not suited to the existing poor-quality pastures. On top of that, El Niño weather phenomena resulted in very low levels of rainfall between late 1997 and early 1998, further diminishing the quality of these pastures. As a result, most animals died during the first two years of activity. The peasants who did not lose their beasts sold them and used the money to survive.

In the wake of this failure, the peasant association was in a very complicated financial position. Its members were collectively liable for the payment of two debts: the 30 per cent tranche that remained to be paid on the estate's purchase and the production loan that had been used to acquire the cattle. The association stopped paying instalments on the loan. Its efforts then became focused on restructuring the debt. Some of the affiliates met with Incora officials and prepared a request to obtain a second production loan intended to fund a palm oil production project, with a commitment to use the entirety of the proceeds to reimburse the two previous debts.

Las Franciscas

Ten kilometres north of Guacamayal, down a straight road running alongside the railway, is the small village of Iberia. Las Franciscas, a 100-hectare estate, is located nearby. Until 1986, it belonged to three banks that had seized the land following a foreclosure procedure. From that time, the ownership of Las Franciscas had been a source of conflict between landlords and landless peasants. In 1987, a local association affiliated with the ANUC national peasant movement was created in Iberia. It brought together peasants who had worked their whole life as agrarian labourers and who were looking to obtain a plot from Incora. They had reasonable chances of success. In the previous decade, the morose economic context had resulted in the purchase by Incora of a number of estates in the area. Las Franciscas was one of these abandoned lands, which had been unoccupied and covered by weeds for years. The peasants then submitted the allocation request to Incora and began farming the land that they expected would become theirs. However, the peasant occupation of the estate corresponded to the precise moment when the local economy was recovering. In 1986, the firm Cacaotera de Orihueca purchased the estate from the banks, but apparently did not resume any production. In late 1987, a man named Antonio Riascos, presumably an associate of Cacaotera, expelled the community from the estate. According to the peasants, he commissioned a local gang, known as the Polvorín, to threaten them and seize the land.

The 1990s booms and busts unfolded, and new prospects ultimately became available to the peasant community. In 1991, the land had been sold to Agrícola Eufemia, a subsidiary of the multinational Dole. Eufemia set up a banana plantation, which was active until early 1996. In March of that year, in the middle of the banana crisis, half of the trees were cut down. Two months later, in May, a storm destroyed the remaining half. Later, in September, a group of FARC guerrillas burned down the farm's packing facility and the adjacent offices. During those years, the members of the peasant community had remained in Iberia, living a few kilometres away from Las Franciscas. Most had found jobs as farm workers, but they had not abandoned their hopes of eventually becoming independent peasant farmers. When Dole's activity ceased and its remaining facilities were destroyed by the FARC, the peasants returned to the land. Twenty families settled on the estate. Most of them built their permanent homes on Las Franciscas and devoted themselves to the production of various food crops, both for their own consumption and for sale in the neighbouring villages.

Following the second occupation of the estate, the leaders of the peasant group reopened the Incora allocation procedure. In May 2000, an official survey was ordered by the institution. The subsequent report established that 47 adults lived on the estate, where they 'farmed surfaces between 1.5 and 3 hectares per family, consisting of fruit-bearing trees, subsistence crops, and vegetables', and living in houses made of clay and straw.[25] Additionally, the report corroborated the fact that that 'the farming corporation Eufemia Ltd. does not exercise or undertake any form of administration of the plots'.[26] This report arguably established the legal basis for the expropriation procedure and the allocation to the peasant community, but there are no further indications that the institution has undertaken additional steps in this respect.

Vulnerability and dispossession

The Las Franciscas community was attacked in September 2001, when the Terán brothers – Jorge, Gustavo, and Miguel – who were prominent members of the community and had been instrumental in the group's decades-old struggle for land, were murdered by a group of paramilitary soldiers. This was an act whose sole purpose was to evict the peasants – who were considered to be squatters – from the estate. However, while the families of the three murdered brothers left the farm after their assassination, most of the other members of the community stayed on the premises. After two and a half years of threats and harassment, a new murder was committed by the paramilitaries in March 2004, that of José Concepción Kelsy, then president of the community's peasant association. When the community refused to leave and a new member took on the role of leader of the association, an attempt on his life was made as well, forcing him to flee the region. In January 2005, community leader Abel Bolaños was shot dead by a group of paramilitary soldiers. This last murder pushed the remaining members of the peasant collective to abandon their farms. A few months later, a banana plantation was established on the estate by the firm Agrícola Eufemia. In 2009, Dole sold all its production operations to focus exclusively on the export and marketing of the produce. Agrícola Eufemia was purchased by one of the largest banana producers in the area and was still in operation at the time of my field research in July 2018.

In Tierra Grata, the community was exposed to paramilitary activities in a different way. Due to the estate's strategic location, it became a paramilitary headquarters. Beginning in early 2001, as the levels of violence were on the rise across northern Magdalena, a paramilitary group associated to AUC began to use the estate to torture their victims and hide their corpses. Years later, in 2007, a police investigation would unearth a mass grave with 15 bodies. In September 2001, one of the members of the association was killed by the militia members; his family fled the estate, fearing for their lives. In the following three years, many others left as well. They were driven away by a combination of terror and poverty, as the failure of the cattle farming project had done away with any prospects of economic development. By 2004, only four families remained on the estate. In May of that year, they were ordered to leave by commander Carlos Tijeras of the AUC.

As with the previously studied cases, further steps were necessary to legalize the situation of both Tierra Grata and Las Franciscas. Here, the legal manoeuvres' success was determined by the ambiguous and fragile status of the peasant's claims. From this point of view, the communities of Las Franciscas and Tierra Grata faced different but equally excruciating challenges.

In Las Franciscas, the peasant community was particularly vulnerable. While all the legal conditions had been fulfilled for the expropriation and subsequent reallocation of the land, Incora had not taken decisive steps to that effect. After the peasants had been expelled by paramilitary forces in 2004, the members of the community were summoned to meet with the landowner's representatives. According to them, the meeting took place on a neighbouring commercial farm belonging to the same corporation, and armed paramilitary militiamen were present. They were told that the company was seizing control over the land and that they were considered as squatters. However, the firm's representatives told them that they would be compensated for their lost crops. Small sums of money, ranging from 50 to 250 US dollars, were given to each peasant, and they were told to sign a receipt.

The legal situation was more complicated in Tierra Grata. When the peasants were expelled by the paramilitaries in 2004, they also received similar compensation for their crops, but the amounts were higher, ranging between 200 and 1600 US dollars. This was not, however, a legal purchase. In parallel to the threats and the forced displacement, the estate was foreclosed by the Agrarian Bank and auctioned-off for 270,000 US dollars. The purchaser was Juan Carlos Serrano, a prosperous palm oil entrepreneur and a member of an influential family. Since the case became public knowledge in the late 2000s, Serrano has always claimed that he had nothing to do with the paramilitary-led eviction, and that he had simply purchased the estate from the bank. The existence of past debts and the apparent legality of the subsequent purchase has hindered the Tierra Grata community's capacity to defend its claims. I will return to this case in the closing chapter of the book, as the challenges faced by this community in 'post-conflict' times exemplify the reproduction of the inequality regime that was consolidated through armed violence.

Discussion and conclusion

These cases of violence and dispossession serve to illustrate the argument that I have sketched out above: all these land-grabbing stories took place in a grey area in which violence is intimately related to market failure, legal obstacles, and labour conflicts. They are a testament to the need to re-embed the analysis of violence within a broader context of exploitation and domination, to study land access and loss not as random, isolated events, but as a process structured by power relations. Within these power relations, violence plays an important role, but it interacts with other forms of exclusion, such as agrarian development policies (or the lack thereof), judiciary activity, or market forces (Hall *et al.* 2011, 4).

How were the stories in La Marcela, Diana María, Las Franciscas, and Tierra Grata structured by power relations? In each of these four cases, the conditions governing peasants' access to land interacted with the violence perpetrated by the paramilitaries, rendering these rural communities particularly vulnerable to dispossession. The cases of La Marcela and Diana María are very similar. In both instances, the peasant's claims were rendered particularly vulnerable by the limitations of the court system. While the employment tribunals did recognize that both peasant collectives had a right to be compensated for unpaid wages and benefits, the occupation of the estates had taken place outside the realm of the law. This occupation was seen by the workers as legitimate compensation for their former bosses' debts, but as a matter of law, it was not legally recognized. As a consequence, once the paramilitary grip over the region had grown in strength, these two communities could be expelled as squatters.

The Las Franciscas community's claim over land was rooted in the long struggle of Colombian peasants for the redistribution of idle estates. The members of this collective held a firm belief that they had the right to occupy the estate by virtue of their long history of attempting to settle down there, which dated back to 1987. However, while the agency in charge of land allocation recognized in 2000 that they met the criteria needed to affirm a legal right to the land, they never obtained ownership rights over the estate. In a manner not unlike that of the communities of workers-cum-peasants of La Marcela and Diana María, they could be considered by the company owning the land as unlawful occupants.

The case of Tierra Grata illustrates a different configuration in which the failure of agrarian development policies and the characteristics of market-driven land reform became an obstacle to the peasants' capacity to remain on land that was legally theirs. The failure of their cattle farming project predated the arrival of paramilitary militias on the estate. The peasants' economic vulnerability, paired with the subsequent bank foreclosure on the estate, transformed the peasants into squatters on their own land. Even more clearly than in the other three cases, their violent eviction from the estate was not a turning point, but the culmination of a process of gradual exclusion.

All three cases illustrate an argument made in the introduction of this book: defining violence as a self-standing reality, disconnected from less spectacular forms of domination, necessarily misses out on an important piece of the picture. My take here has been the opposite, as I contend that wartime plundering must be embedded in a broader analysis of the political economy, understood as the articulation between accumulation, institutions, power, and identities (Marchal and Messiant 1997).

This conclusion has far-reaching consequences for the study of periods considered as 'post-conflict transitions'. If violent dispossession is the most radical embodiment of a broader inequality regime, then how should it be studied once its proceeds are no longer seized through violent means, but transformed into legitimate accumulation? If the relationship between wartime and peacetime economy is one of continuity rather than rupture, then how should we interpret the efforts of post-conflict practitioners to trace clear-cut lines between war and its aftermath?

The coming chapters will present the various political and analytical challenges embodied by these questions. Beginning in the second half of the 2000s, the association between land and conflict became a common theme among peacebuilding practitioners, both in Colombian institutions and internationally. This led to the passage of a Victims Act in 2011, which made land restitution and return possible for dispossessed peasants. Such a context was not unrelated to the peace negotiations that took place between the Colombian government and the FARC between 2012 and 2016. Overall, the promise to put an end to the cycle of violence and dispossession was seemingly becoming the foundation for a new social contract.

From the point of view of most spectators, these progressive policies were at the opposite end of the spectrum from past histories of violence. In a certain way, this was true, as they were promoted by people who strongly believed in the state's capacity to transform history and overcome the legacies of violence and exploitation. However, the implementation of most of these policies ended up – despite the best wishes of their promoters and designers – consolidating a proprietarian social order that had been built in the commotion of war. This incongruity, which is overlooked by more simplistic interpretations of the relations between war, peacebuilding, and capitalism, will be the subject of the remainder of this book.

Notes

1 See, among others: Machado and Meertens (2010), Restrepo Echeverri and Franco Restrepo (2011), Baquero Melo (2014), Rodríguez González (2014), Gómez *et al.* (2015), Osorio Pérez (2015), Mercado Vega (2017), Quinche Roa *et al.* (2018).

2 As suggested, among others, by Uribe-Kaffure (2014), Ojeda *et al.* (2015), Ojeda (2016), García Arboleda (2019), Ballvé (2020).

3 Similar approaches include: Gutiérrez and Vargas (2016), Counter (2019), Gutiérrez *et al.* (2019).

4 This approach is strongly influenced by the important book by Hall, Hirsch, and Li (2011).

5 The four case studies presented in the chapter are based on various oral and written sources. In July 2018, I interviewed the leaders of the communities of Diana María and Tierra Grata (interviews in Santa Marta and Guacamayal), and I was fortunate enough to

be able to meet with most of the surviving members of Las Franciscas community (focus group and interviews in Iberia). In Santa Marta and Bogota, I met with the attorneys defending the three aforementioned cases and the case of La Marcela. In Santa Marta, I interviewed several officials from the land restitution unit who had been active in the four cases. In addition to these interviews, I had access to the files for the four restitution demands; they are registered in the following jurisdictions: Juzgado segundo civil del circuito especializado en restitución de tierras de Santa Marta (land restitution court), case no. 2017-00067 (La Marcela); Juzgado segundo civil del circuito especializado en restitución de tierras de Santa Marta, case no. 2016-0048 (Diana María); Tribunal superior de Cartagena, Sala de restitución de tierras (chamber of land restitution, high tribunal of Cartagena) case no. 2014-0009 (Las Franciscas); UAEGRTD (land restitution unit), Resolución RM00925, 7 December 2015 (Tierra Grata).

6 The numbers from the banana district for the entire 1990–2003 period concern the municipalities of Aracataca, Ciénaga, El Retén, Fundación, Guamal, Sitionuevo, and Zona Bananera (from its creation in 1999)

7 All cases of paramilitary murders are drawn from CINEP's (2004) database on political violence.

8 Juzgado Único Especializado [Special Criminal Court], Santa Marta, 24–27 March 2009.

9 Interview, Ciénaga, March 2009.

10 Interview, Ciénaga, March 2009.

11 Quoted in Corte Constitucional, decision No. T-952, 17 October 2003, p. 4.

12 Interview, Ciénaga, March 2009.

13 UAEGRTD (land restitution unit), Resolución RM00925, 7 December 2015, p. 60.

14 Ejército Nacional – 20° Brigada, Batallón de Inteligencia no. 1. Santa Marta, 8 May 1998. 001830/RR20-BITE-1-INT 10 252. Anotaciones de inteligencia sobre Adán, Rigoberto y Camilo Rojas. Private archives.

15 Chiquita in-house counsel handwritten notes about a front company set up by paramilitaries in Santa Marta to collect security payments from Banadex, 6 March 2000. Included in Evans (2011).

16 Tribunal Superior de Bogotá, José Gregorio Mangones Lugo, No. 2007-82791, 31 July 2015, p. 626.

17 Instituto Nacional Penitenciario y Carcelario (National Correctional Agency), Declaration of José Gregorio Mangones Lugo, Barranquilla, 29 October 2009, pp. 4–5. Private archives.

18 European Currency Unit; one ECU was later valued at 1 euro.

19 The municipality of Zona Bananera was created in 1999 at the heart of the banana district and covers a territory that formerly belonged to the municipality of Ciénaga.

20 UAEGRTD, Solicitud colectiva de restitución jurídica y material (land restitution request), La Marcela, p. 145. File location referenced above.

21 UAEGRTD, Declaration of Carlos Urueta Julio, Diana María, book no. 2. File location referenced above.

22 UAEGRTD, Solicitud colectiva de restitución jurídica y material (land restitution request), Diana María, p. 58. File location referenced above.

23 *Ibid.*, p. 59

24 Interview, Guacamayal, July 2018.

25 Tribunal superior de Cartagena, Sala de restitución de tierras (chamber of land restitution, high tribunal of Cartagena), decision no. 2014-0009, 24 January 2018, p. 74–5.

26 *Ibid.* p. 73.

References

Ballvé, T. 2020. *The Frontier Effect: State Formation and Violence in Colombia.* Ithaca, NY: Cornell University Press.

Baquero Melo, J. 2014. *Layered Inequalities: Land Grabbing, Collective Land Rights and Afro-Descendant Resistance in Colombia.* Berlin: LIT Verlag.

Bonet, J. 2000. *Las exportaciones colombianas de banano.* Cartagena: Banco de la República.

Botero, F. 1990. *Urabá: colonización, violencia y crísis del Estado.* Medellín: Universidad de Antioquia.

Bucheli, M. 2005. *Bananas and Business: The United Fruit Company in Colombia, 1899-2000.* New York, NY: NYU Press.

Carroll, L.A. 2011. *Violent Democratization: Social Movements, Elites, and Politics in Colombia's Rural War Zones, 1984-2008.* Notre Dame, IN: University of Notre Dame Press.

CINEP, (Centro de Investigación y Educación Popular). 2004. *Deuda con la humanidad, paramilitarismo de Estado en Colombia.* Bogotá: CINEP.

CORPES, (Consejo Regional de Planificación de la Costa Atlántica). 1990. *Orden social en la región caribe y el proceso de paz.* Cartagena: CORPES.

Counter, M. 2019. In Good Faith: Land Grabbing, Legal Dispossession, and Land Restitution in Colombia. *Journal of Latin American Geography*, 18(1), 169–92.

DANE, (Departamento Administrativo Nacional de Estadística). 2016. *Tercer censo nacional agropecuario.* Bogotá: DANE.

Deininger, K. 1999. Making Negotiated Land Reform Work: Initial Experience from Colombia, Brazil and South Africa. *World Development*, 27(4), 651–72.

El Heraldo. 2011. Candidaturas en el Magdalena: ¿qué pasó con la renovación política? *El Heraldo*, 18 Jun.

El Tiempo. 1991a. Ciénaga: llegó la hora buena del banano. *El Tiempo*, 21 Oct.

El Tiempo. 1991b. Ciénaga, entre la euforia y el miedo. *El Tiempo*, 7 Oct.

El Tiempo. 1997. Autodefensas asesinan cuatro personas en Zona Bananera. *El Tiempo*, 29 Aug.

Evans, M. 2011. The Chiquita Papers. *National Security Archive*, 7 Apr.

García, C.I. 1996. *Urabá: región actores y conflicto, 1960-1990.* Bogotá: INER-Cerec.

García Arboleda, J.F. 2019. *El exterminio de la isla de Papayal: Etnografías sobre el Estado y la construcción de paz en Colombia.* Bogotá: Editorial Pontificia Universidad Javeriana.García, C.I. 1996. *Urabá: región actores y conflicto, 1960-1990.* Bogotá: INER-Cerec.

García-Echeverría, L. 2019. *La economía colombiana y la economía mundial. 1950-2017.* Bogota: Pontificia Universidad Javeriana.

Gill, L. 2016. *A Century of Violence in a Red City: Popular Struggle, Counterinsurgency, and Human Rights in Colombia.* Durham, NC: Duke University Press.

Gómez, C.J.L., L. Sánchez-Ayala, and G.A. Vargas. 2015. Armed Conflict, Land Grabs and Primitive Accumulation in Colombia: Micro Processes, Macro Trends and the Puzzles in Between. *The Journal of Peasant Studies*, 45(2), 255–74.

Grajales, J. 2011. The Rifle and the Title: Paramilitary Violence, Land Grab and Land Control in Colombia. *The Journal of Peasant Studies*, 38(4), 771–92.

Grajales, J. 2013. State Involvement, Land Grabbing and Counter-Insurgency in Colombia. *Development and Change*, 44(2), 211–32.

Grajales, J. 2015. Land Grabbing, Legal Contention and Institutional Change in Colombia. *The Journal of Peasant Studies*, 42(3–4), 541–60.

Gutiérrez, F. and J. Vargas. 2016. *El despojo paramilitar y su variación: quiénes, cómo, por qué.* Bogotá: Universidad del Rosario.

Gutiérrez, F., R. del P. Peña-Huertas, and M.M. Parada-Hernández. 2019. *La tierra prometida: Balance de la política de restitución de tierras en Colombia.* Bogotá: Editorial

Universidad del Rosario.Gutiérrez, F. and J. Vargas. 2016. *El despojo paramilitar y su variación: quiénes, cómo, por qué*. Bogotá: Universidad del Rosario.

Hall, D., P. Hirsch, and T.M. Li. 2011. *Powers of Exclusion: Land Dilemmas in Southeast Asia*. Honolulu, HI: University of Hawaii Press.

Josling, T.E. and T.G. Taylor, eds. 2003. *Banana Wars. The Anatomy of a Trade Dispute*. Cambridge, MA: CABI.

Lombana, J. 2012. *Desarrollos y Estructuras Del Mercado Del Banano de Exportación En Colombia (1995–2010)*. Barranquilla: Universidad del Norte.

Londoño Botero, R., ed. 2016. *Tierras y conflictos rurales. Historia, políticas agrarias y protagonistas*. Bogotá: Centro Nacional de Memoria Histórica.

Machado, A. 1999. *El mercado de tierras en Colombia: una alternativa viable?* Bogotá: Tercer Mundo.

Machado, A. and D. Meertens, eds. 2010. *La tierra en disputa. Memorias de despojo y resistencia en la Costa Caribe (1960-2010)*. Bogotá: Centro Nacional de Memoria Histórica.

MADR, (Ministerio de Agricultura y Desarrollo Rural). 2020. *Evaluaciones Agropecuarias Municipales. Oficina de planeacion y prospectiva*. Bogotá: MADR.

Marchal, R. and C. Messiant. 1997. *Les Chemins de la guerre et de la paix. Fins de conflit en Afrique orientale et australe*. Paris: Karthala.

Martínez, L. and W. Renán. 2011. Memorias y narrativas: una reconstrucción del contexto de violencia en el Magdalena (1980-2009). *In*: F. Silva, ed. *Reconstrucción de la memoria oral de los desmovilizados y desplazados en los departamentos del Magdalena, Cesar y Guajira entre 1980 y el 2009*. Santa Marta: Oraloteca. Universidad del Magdalena, pp. 358–531.

Mercado Vega, A.J. 2017. Contrarreforma agraria y conflicto armado: Abandono y despojo de tierras en los Montes de María, 1996–2016. *Economía & Región*, 11(2), 197–248.

ODH, (Observatorio de derechos humanos de la vicepresidencia de la República de Colombia). 2014. *Estadísticas de violaciones a derechos humanos y Derecho Internacional Humanitario (1990-2013)*. Bogotá: Vicepresidencia de la República.

Ojeda, D. 2016. Los paisajes del despojo: propuestas para un análisis desde las reconfiguraciones socioespaciales. *Revista Colombiana de Antropología*, 52(2), 19–43.

Ojeda, D., J. Petzl, C. Quiroga, A.C. Rodríguez, and J.G. Rojas. 2015. Paisajes del despojo cotidiano: acaparamiento de tierra y agua en Montes de María, Colombia. *Revista de Estudios Sociales*, (35), 107–19.

Ortíz Sarmiento, C.M. 2007. *Urabá: pulsiones de vida y desafíos de muerte*. Bogotá: La Carreta Social.

Osorio Pérez, F.E. 2015. Tramas entre paramilitarismo y palmicultura en Colombia. *Memoria y Sociedad*, 19(39), 11–28.

Quinche Roa, J.M., P. Perdomo Vaca, and J. Vargas Reina. 2018. *El despojo paramilitar en el Magdalena: el papel de las élites económicas y políticas*. Bogotá: Observatorio de Restitución y Regulación de Derechos de Propiedad Agraria.

Renán, W. 2007. *Contextualización del desplazamiento en el departamento del Magdalena y Santa Marta (1997-2007)*. Santa Marta: Universidad del Magdalena, Vicerrectoría de Investigación, Fonciencias, Sección I, informe final convocatoria 2004.

Restrepo Echeverri, J.D. and V.L. Franco Restrepo. 2011. La toma de la tierra: lógicas de guerra y acumulación en el Bajo Atrato. *In*: M. Romero, ed. *La economía de los paramilitares*. Bogotá: Random House Mondatori, pp. 269–410.

Rodríguez González, I. 2014. Dispossession, Wastelands and Armed Conflict in Puerto Gaitán and Mapiripán (Meta, Colombia) between 1980 and 2010. *Estudios Socio-Jurídicos*, 16(1), 289–314.

Romero-Prieto, J.E. 2007. Movilidad social, educación y empleo: los retos de la política económica en el Departamento del Magdalena. *In*: A. Meisel, ed. *Las economías departamentales del Caribe continental colombiano*. Bogotá: Banco de la República, pp. 422–88.

Suárez, A.F. 2007. *Identidades políticas y exterminio recíproco. Masacres y guerras en Urabá (1991–2001)*. Bogotá: La Carreta.

Tangermann, S. 2003a. European Interests in the Banana Market. *In*: T.E. Josling and T.G. Taylor, eds. *Banana Wars*. Wallingford: CABI, pp. 17–44.

Tangermann, S. 2003b. The European Common Banana Policy. *In*: T.E. Josling and T.G. Taylor, eds. *Banana Wars*. Wallingford: CABI, pp. 45–66.

Uribe-Kaffure, S. 2014. Transformaciones de tenencia y uso de la tierra en zonas del ámbito rural colombiano afectadas por el conflicto armado. El caso de Tibú, Norte de Santander (2000-2010). *Estudios Socio-Jurídicos*, 16(1), 243–83.

Viloria de la Hoz, J. 2008. *Banano y revaluación en el departamento del Magdalena (1997-2007)*. Cartagena: Banco de la República.

4

FORCED DISPLACEMENT, HUMANITARIAN EMERGENCY, AND LAND DISPOSSESSION

From the middle of the 1990s, people like those who were expelled from Las Franciscas or La Marcela joined the incessant streams of internal exiles, as Colombia was awarded the unenviable status of the world's second-worst case of internal displacement, just behind Sudan. These thousands, then millions of internally displaced persons (IDPs) added a further layer of complexity to the already intractable equation of land and war. The struggle faced by many was precisely that of untangling this complexity. Within human rights organizations, development and humanitarian assistance agencies, social movements, and state institutions, multiple efforts were made in attempts to push for the recognition of the close ties between land and war. The champions of these efforts were activists, politicians, intellectuals, and social leaders. Many of them shared the deep conviction that land tensions were one of the primary sources of violence, and that no peacebuilding effort could be successful as long as the issue of rural property remained unresolved.

In 2011, President Juan Manuel Santos ratified the Victims Act. It was, according to him, the most revolutionary piece of legislation in recent years. The conviction that this law constituted a turning point in the history of the Colombian war was widely shared in policymaking circles and the media. The historic nature of this success was acknowledged by then UN Secretary General Ban Ki Moon, who travelled to Bogota to celebrate the enactment of the law. The text was indeed revolutionary. It granted a set of rights to the victims of the war, and among them was a right to land restitution. Acknowledging that armed conflict had resulted in a massive process of land grabbing and spoliation, the law created an extraordinary restitution procedure. Moreover, the law was a promise for the millions of victims of land dispossession in Colombia: the promise that the state would not simply leave the restitution problem in the hands of judges and courts, but would support claimants' efforts at every step of the process, knowing that for years the law had been instrumentalized by the powerful against the powerless. It was an inspiring promise for impoverished peasants like those of Diana María or Tierra Grata, whose dreams had been shattered by paramilitary militias.

For many activists, intellectuals, and policymakers, this law was a victory earned through years of political struggle. The agrarian roots of the conflict had finally been recognized and addressed. In retrospect, this feeling was strengthened by the start of formal peace negotiations in Havana with the FARC in 2012, the agrarian chapter agreement of 2013, and by the final peace accord of 2016. Years after the enactment of the 2011 law, the Victims Act was seen by many – including those who negotiated the peace agreement – as the legislation that paved the way for the peace talks.

My objective here is to provide a more nuanced view that reinscribes this genealogy within the broader context of peace and land policies. The importance of restitution legislation cannot be understated, even if its implementation has been incomplete at best (Gutiérrez *et al.* 2019). It is true that the 2011 law scored an ideological victory, as it became widely acknowledged that land conflicts were a critical element in the unfolding of war. From that point of view, the emergence of agrarian issues as a central topic of the post-conflict agenda marks the success of many people's years-long political struggle. This chapter will provide a brief account of those efforts. However, it will also defend an alternative view of this process of agenda setting. My argument is that the price to pay for the emergence of the land issue as a peacebuilding problem was to reduce its scope. True, the 2011 law was a revolutionary piece of legislation, as J.M. Santos eloquently said, but it was only revolutionary for certain victims of land spoliation. In a certain way, as the law recognized that land and war were inseparably linked, it reduced this relationship to a single dimension: dispossession. By the same token, it separated violent land grabbing from its economic and political substrate. The law made it an extraordinary manifestation of extraordinary events. It traced a fix line between criminal spoliation and legitimate – even if outright – inequality.

I will proceed in three steps. First, I will trace the emergence of forced displacement as a public problem, in the middle of one of the largest humanitarian emergencies of the contemporary Americas. Second, I will show how internal displacement was connected to land dispossession. Last, I will show that, while great ambitions surrounded the restitution policy, its scope was carefully circumscribed to preclude any attempt to open up the redistributive agenda.

Internal displacement: from dark to light

Years before the recognition of the close ties between land and armed conflict, human rights activists and humanitarian practitioners had confronted the state in hopes of obtaining recognition for the growing population of IDPs, the majority of whom were rural dwellers. I will briefly sketch here the convergence of different forms of activism, international influence, and political struggles around the issue of forced displacement.

Turning a blind eye

In spite of the fact that an estimated 600,000 people were forcibly displaced between 1985 and 1995, the Colombian government did not recognize internal displacement as a problem until the middle of the 1990s. Rural migration was seen as a by-product of violence, but also as a natural phenomenon consistent with the modernization of society, the crisis of peasant agriculture, and the decline of rural economies (Vidal López 2007; Aparicio 2009). The emergence of the problem in the agenda, and the construction of IDPs as a group endowed with specific rights, ensued from the conjunction of domestic and international politics.

At the international level, internal displacement had begun to transform into an issue of policy attention since the beginning of the 1990s (Cohen and Deng 1998; Weiss and Korn 2006). Refugee status originally applied exclusively to international refugees. While the idea that the Geneva convention could be extended to cover IDPs had been advocated in UN arenas since the 1950s, a more restrictive appraisal had been successfully defended by most major powers. However, norms and standards at the regional level, such as those of the Organization of African Unity or the Inter-American human rights system, came to recognize the urgency of granting this recognition to IDPs (Loescher 2001; Vidal López 2007). In the late 1980s, there was enhanced mobilization in favour if this. Two international conferences – Oslo in 1988 and Guatemala City in 1989 – contributed to defining internal displacement as an international problem. In 1992, UN Secretary-General Boutros Boutros-Ghali appointed Francis Deng, a Sudanese scholar and diplomat, as his Special Representative on Internally Displaced Persons. Deng visited Colombia in 1994, and expressed serious criticisms about the government's inability to provide an accurate assessment of the internal displacement problem (Aparicio 2009). Internationally, this course of events represented a shift for international law, as it opened the door to far more comprehensive intervention from the international community in the internal affairs of sovereign states (Loescher 2001; Weiss and Korn 2006; Cohen 2012).

In the meantime, the Colombian government was opposed to seeing this international perspective applied to its own domestic situation. The Gaviria administration's official policy was to politely refuse any attempt by international actors – such as the UN High Commissioner for Refugees or the Permanent Consultation on Internal Displacement in the Americas – to intervene in its internal affairs (Vidal López 2007, 139–42). However, beginning in the mid-1980s, a vigorous mobilization of civil society actors took place. Dávila (2018) distinguishes two trends of collective action. The first came from the Catholic Church, which was one of the main actors in the area of humanitarian assistance and had been particularly exposed to the growing number of IDPs. In 1985, the Bishop's conference commissioned a small group of social scientists, many of them with close ties to the Jesuits, to conduct surveys on the internal exodus (Dávila 2018, 159). During the same period, the concern for internal displacement grew among the human rights and political left circles. As activists themselves became the primary target of paramilitary and death-squad violence, many of them were victims of forced displacement (Tate 2007). This experience fed the emergence of IDPs as a political cause, in a context in which loose networks of activists were transforming into much more professionalized and internationally connected NGOs. In 1992, activists from these circles created the first NGO specifically devoted to the cause of IDPs, CODHES (Consultancy on Human Rights and Forced Displacement) (Dávila 2018, 163).

Rhetorical commitment and constitutional review

The arrival of a new administration in August 1994 offered a favourable context for the conjunction of international agenda-setting and domestic mobilization. That year, Ernesto Samper was elected to the presidency with a social-democratic agenda and the promise to grant a larger place to human rights in security policies. However, at a very early juncture, his administration was undermined by credible evidence tying his campaign management to powerful Cali drug traffickers (Leal Buitrago 1996). Stigmatized on the international stage by the illicit campaign financing, Samper attempted to use the human rights agenda as a way of gaining some respect and credibility (Daviaud 2010).

Early in 1995, the Bishop's conference released the first nationwide assessment of internal displacement (Dávila 2018, 173). It was the result of a decade's work and enjoyed the moral authority of the Catholic Church. The release of the report coincided with the media visibility of the problem of forced migration, as thousands of people were leaving the region of Urabá due to violent attacks by paramilitary and guerrilla groups. The government responded by endorsing most of the document's conclusions; a few months later, the first policy paper intended to create an assistance policy for IDPs was released (DNP 1995). The document brought together both national and international sources of policy, including many of Francis Deng's conclusions, the definition of IDPs by the Permanent Consultation on Internal Displacement in the Americas, and the conclusions reached by the Bishop's conference (Cardona-Fox 2019, 97). It marked the official recognition of the existence of internal displacement as a public problem.

The following years were characterized by a classic situation of 'rhetorical commitment', i.e., a succession of official decisions and announcements with very little concrete impact (Cardona-Fox 2019). A new policy paper was released in 1997 recognizing the administration's failure to protect IDPs and attributing it to a lack of coordination, managerial shortcomings, and financial drawbacks (DNP 1997). A few months later, an ambitious and comprehensive law on internal displacement was passed by Congress, but the government did not seem committed to applying it, as several years passed before the implementing decrees were ultimately issued. Yet another policy paper was published in 1999; it recognized that the already identified coordination problems had not been addressed, the application of the law was incomplete at best, and the state was unable to monitor both the magnitude of the internal displacement problem and the impact of its policies (DNP

1999). That same year, a new visit by Francis Deng reached similar conclusions (Rodriguez Garavito and Rodriguez Franco 2010, 95).

Throughout these years, the Colombian government avoided further scrutiny from the international community by conveying an image of commitment and good faith (Garibay and Guerrero-Bernal 2007; Daviaud 2010). In 1997, the government officially requested the assistance of the UN High Commissioner for Refugees (UNHCR), whose role was initially limited to policy advice. In 1999, the UNHCR's scope of intervention was widened, in order to include direct cooperation in the implementation of assistance programmes (Vidal López 2007, 162). Colombia's cooperative attitude brought about benefits for both the government and the UNHCR's Bogota office. According to Aparicio (2009, 183), the UN body consistently avoided adopting an oppositional stance, seeking to become a source of policy advice and best practices, a stance that was seen in Geneva as an example to follow. Even if the government generally failed to implement the policy that it had officially adopted, international agencies like the UNHCR refrained from openly criticizing it in order to avoid antagonizing a partner that was seen as responsible and cooperative.

The most important transformation in Colombia's IDP policy, which forced the government to move from rhetorical commitment to concrete action, was the intervention of the Constitutional Court in 2004. In a much-commented decision, the judges ruled that IDPs' dreadful living conditions were 'accompanied and aggravated by structural failures in state response', thus constituting an 'unconstitutional state of affairs' (Rodriguez Garavito and Rodriguez Franco 2010, 45). The constitutional nature of the ruling enabled the court to move from the judgement of a certain number of cases to the evaluation of a whole area of public policy. This kind of judicial activism reflects an emerging trend in Latin America and other regions of the Global South in which constitutional review is used as a tool for social advancement (Gauri and Brinks 2010; Rodriguez Garavito 2014). One salient feature of the ruling was the monitoring procedure established by the Court; through 84 follow-up decisions and 14 public hearings, the judges evaluated the government's progress in meeting the objectives set by the Court, putting unprecedented pressure on public officials to take concrete steps toward actually implementing the ruling.

According to Rodríguez Garavito and Rodríguez Franco (2010, 50–63), the ruling had two sets of political consequences. First, it framed the problem of forced displacement in a novel way. Assistance to IDPs had previously been construed as a matter of social solidarity; now, the Court had defined it as a basic human right. Second, the Court reshaped the policy field of assistance to IDPs. Not only did the work of constitutional monitoring place the Court at the centre of a broad policy network, but it also established the responsibilities and the resources available to other actors. According to the ruling, a range of public agencies should be held accountable for the response (or the lack thereof) to IDPs' needs. The ruling also empowered public monitoring entities, such as the Inspector's Office (*Procuraduría*) and the National Accounting Office (*Contraloría*) to participate in the policy's evaluation. Lastly, it provided NGOs and social movements the legal resources needed to hold the state accountable. In parallel to the Court's action, a monitoring committee was set up in 2005 by a group of public intellectuals, NGOs, and scholars from Los Andes University, one of the country's premier private establishments. The Court relied on this monitoring committee, which was depicted as a 'civil society body', to produce regular surveys on different aspects of the IDPs' living conditions and the state response to the humanitarian emergency.

From displacement to dispossession

Very early in the history of the emergence and transformation of the internal displacement problem, the link with land dispossession was made evident by domestic and international actors. Land grabbing was identified as both a motive for displacement and a consequence of the dreadful living conditions of peasants in the countryside. However, as with IDPs, a long struggle was necessary before the government would recognize the problem, and an even more challenging struggle would be required before state agencies would display a real commitment to the daunting task of providing a safeguard procedure intended to prevent massive land grabs. This was particularly true because, from 2002 to 2010, the political arena was dominated by hard-line, right-wing president Álvaro Uribe and his supporters, for whom any discussion about victims, reparation, and restitution equated to aligning with 'terrorist' guerrillas.

The emergence of dispossession

According to Dávila (2018, 180), the first known inquiries into the assets abandoned by IDPs dated from 1996 and 1998 and were undertaken by CODHES, which had already become the flagship NGO for matters of internal displacement. During that period, the idea that land dispossession was not simply a consequence of forced displacement, but rather a source of motivation for spurring people to flee, made its way into academic circles. In 1999, Absalón Machado, one of the leading scholars in the area of Colombian agrarian studies, used cadastral data to trace the process of land concentration in zones where paramilitary groups held sway; his argument was that land accumulation was becoming a primary driving force behind the expansion of paramilitary militias and the forced exile of hundreds of thousands of peasants (Machado 1999, 203–8). These activist and scholarly analyses found an echo among the international development and human rights establishment. In 1999, for instance, the Inter-American Commission on Human Rights considered that 'concealed behind the phenomena of violence and armed confrontation, are economic interests associated with the so-called agrarian counter-reform that affects small and medium-scale landowners' (IACHR 1999, ch. 6).

The idea of an intimate link between forced displacement and dispossession led to new efforts to measure the magnitude of dispossession. In the absence of aggregate data on the land and assets lost, the method was generally to measure, across a sample of IDPs, the percentage of people who were peasants, as well as the average surface area of their farm, and to extrapolate these figures to the general population of IDPs. In 2000, CODHES announced an estimate of 1.7 million hectares abandoned by IDPs between 1995 and 1999 (Dávila 2018, 184). In 2001, the World Food Programme made a much higher estimate – 4 million hectares – but provided no details about the methods used for the calculation (WFP 2001). All these estimates conflated land abandonment and dispossession. In the absence of better measurement tools, and considering the powerful agrarian interests involved in the war, it was probably a good guess to consider that IDPs' farms would seldom remain unoccupied. In parallel to those institutional mobilizations, grassroots activists were also striving to shed light on the dreadful situation of entire communities who had been evicted from their lands and stripped of their livelihoods. Many of these activists relied on international alliances – sometimes rendered possible by the support of Colombian NGOs – to attract the attention of the media and the government (Allain 2019). In many cases, these mobilizations also relied on previous networks, such as those of peasant organizations, leftist parties, and workers' unions (Celis 2018).

While discussion regarding the topic of dispossession was becoming widespread among members of civil society and international aid circles, state institutions were much less reactive. The 1997 law on internal displacement had created a procedure intended to protect IDPs' assets, but in the absence of a decree establishing the concrete measures to be followed, the law remained ineffectual. Studies by Estrada and Rodríguez (2014, 2016), as well as by Dávila (2018), trace the process of mobilization within the state, spearheaded by a group of committed officials in an attempt to bring the legal provisions to life. These were professionals working for the *Red de Solidaridad Social* (RSS), an official agency in charge of assistance to IDPs, as well as for the Colombian Institute for Agrarian Reform (Incora). With the support of the UNHCR and the World Bank, who financed a series of preliminary studies and consultancies, they drafted the decree and lobbied the government, which ultimately led to it being issued in September 2001.

In the following years, this group of officials formed a policy network that occupied a singular place within the state. The same officials who lobbied for the decree set up a grant proposal in response to a World Bank call for projects. Funding was approved by the Bank in 2002, leading to the creation of a 'Project for the Protection of the Land and Patrimony of Displaced Populations' (PPTP) (Dávila 2018, 239). These were peculiar times for the implementation of such a policy. In 2002, the arrival of Álvaro Uribe to the presidency marked a hard-line turn in security policy. Human rights and humanitarian action were readily labelled as anti-state (and thus, pro-insurgent) activities. However, PPTP officials were successful in keeping a low profile all while becoming a 'donor's darling',[1] enjoying the support of powerful development actors such as the World Bank, the EU, Canada, and Sweden. In addition, as the Constitutional Court pursued its close monitoring of the policy on forced displacement, the PPTP's results were used by the government as proof of its commitment to fulfilling the standards set by the court's ruling (Estrada and

Rodríguez 2016, 390–6). Over the following years, the PPTP played a critical role in setting the agenda on forced displacement and land dispossession. The project did not have any authority to implement a restitution policy, but it still endeavoured to prevent the legalization of land grabbing and to document the mechanisms usually used to accumulate land. They implemented a protection policy, including a moratorium on transactions in areas characterized by massive forced displacement (Estrada and Rodríguez 2014, 91). Through aggregate data and field research, they accumulated colossal amounts of information on the multiple links between forced displacement and land accumulation. They identified paramilitary militias as a key actor in land dispossession, as their origins and development were entangled with agribusiness and cattle ranching milieux and their strongholds were located in regions of historical land struggles. In addition, PPTP studies accurately established that land dispossession was not taking place extralegally, but that institutions and legal procedures had been extensively used to launder the grabbed assets (Grajales 2011a; Ballvé 2013).

Transitional justice and victims' rights

In 2003, the government announced that it had reached an agreement under which most of the country's paramilitary militias would be demobilized (Carranza-Franco 2019). The process was fraught with inconsistencies regarding the number and identity of demobilized militiamen and doubts as to the sincerity of their commitment to disarmament (Human Rights Watch 2005). It was also the target of domestic and international criticism on the government's benevolent attitude towards pro-systemic actors that upheld extensive collusive links with influential members of the political elite (Cubides 2006; Grajales 2011b, 2016). However, the fabrication of a legal framework for paramilitary fighters established the conditions needed to import and transplant the principles of transitional justice (Uprimny *et al.* 2006).

By the mid-2000s, transitional justice was becoming a dominant doctrine for the articulation of peacebuilding and justice at the global level (Elster 2004; Lefranc 2009; Hayner 2010). Delphine Lecombe (2014) carefully analyses how the Colombian government's early strategy was to adopt a rhetorical commitment to the principles of truth, reparation, non-recurrence, and reconciliation, while limiting the concrete impact on institutions and policies. However, once these debates, standards, and legal tools began to be included in the legal debate, they were used by multiple actors (NGOs, think tanks, political activists, etc.) to force the government into more binding policies. The Constitutional Court also played an important role, as it modified the legislation passed in 2005 for the demobilization of paramilitary groups, introducing enhanced guarantees for victims and a higher standard of truth for perpetrators.

By 2006, when the Court issued its ruling, the problem of land dispossession was beginning to converge with the debate on transitional justice. On the one hand, as paramilitary groups entered the transitional justice process and the possibilities for inquiring about their strategies for controlling and ruling territories broadened, the intimate connection between violence and land grabbing became even more evident than before (Salinas Abdala and Zarama Santacruz 2012). On the other hand, the law had not established a clear procedure for reparations for lost land. The 2005 Justice and Peace Act had created a National Commission for Reparation and Reconciliation (CNRR). This commission had the task of instituting an 'asset restitution' programme, but enjoyed very little political support (Sánchez 2017, 169). As a result, the mechanisms established by the law – the voluntary contribution or seizure of former paramilitary groups' assets – were acutely ineffective. By 2011, only 17 former members of paramilitary groups had delivered assets for a total value of 5311 million pesos (approximately 2.8 million US dollars) (Carranza *et al.* 2019, 48).

In October 2007, the Liberal Party, which was then moving into a more oppositional stance with regard to Uribe's government, submitted a bill on restorative justice for victims of the armed conflict. The initial draft provided for a general appraisal of IDPs' assets, but no specific approach on land (Dávila 2018). Several rounds of negotiation unfolded, as Liberal Senator Juan Fernando Cristo, who was the law's rapporteur and main supporter, attempted to find a compromise with the government (Cristo 2012). As the PPTP was the only state office with expertise on the matter, officials from the project were called in to contribute to the bill's draft (Dávila 2018). In May 2008, a considerably evolved version of the bill included clear and ambitious provisions in favour of rural land restitution. Most importantly, it introduced what lawyers would come to call 'transitional civil law'. This broad concept referred to the fact that commercial transactions during times of extreme violence and disorder could not be considered as having been undertaken by individuals making use of their free will. Consequently, the state could legitimately revise those transactions to establish which of them had been made under duress (Sánchez 2017). More specifically, this meant that the state could determine the limits of so-called 'outlaw zones', i.e., places where, during a specific period, all real estate transactions were suspected of having been made under duress. Similarly, transitional civil law included the idea of a 'reversal of the burden of proof', which meant that, contrary to a sacrosanct principle of private law, claimants did not have to provide the evidence to support their claim; either defendants or the judge had to support or contradict the claim (Dávila 2018, 292).

According to Dávila (2018, 291-sq), PPTP lawyers promoted these innovations based on three different sources: first, the PPTP pilot programmes implemented in prior years had demonstrated that the law was not protecting the weak but was being used to consolidate land dispossession and accumulation; second, the PPTP had conducted abundant research on various foreign examples and had organized workshops with foreign experts presenting their own experience with land restitution; third, there was a precedent of a restitution policy in contemporary Colombia, implemented with very limited success after 1959 (see also Karl 2017).

Paradoxically, the government's opposition to the Liberal Party's bill did not focus on the principles of transitional civil law, which were arguably its most revolutionary aspects, but on a set of highly political and symbolic provisions of the law, such as the one providing that the state had a responsibility towards victims (for its action or its lack thereof), or the one granting the same treatment to the victims of armed groups and state agents. The principles of transitional civil law seemed to have become mainstream; when the government presented its own version of the bill, the text kept most of the provisions that just a few years earlier would have been perceived as highly disruptive. Observers at the time attributed this to the fact that the administration anticipated a purely rhetorical adherence (Saffon 2010), which was consistent with its attitude on many other issues linked with victims and human rights. However, a closer, sociological examination of policy-making circles in Bogota demonstrates that this attitude was also tied to the fact that the principles of transitional justice had become widely influential throughout the legal community, and that there was not a clear-cut division between the legal professionals advising the government and the Liberal Party (Lecombe 2014; Dávila 2018). Compellingly, similar ideas about transitional civil law were present in other institutional arenas, such as the National Reparation and Restitution Commission, which delivered its guidelines for a reparation policy in July 2010 (Uprimny and Sánchez 2010, 313).

The bill was finally abandoned in June 2009, as president Uribe mobilized his allies in the Senate, arguing that the law was fiscally unsustainable and both legally and morally flawed, as it equated victims of state agents to those of illegal armed groups (Carranza *et al.* 2019, 50). However, the entire legislative debate that unfolded between 2007 and 2009 established the terms of the debate around the extraordinary measures required to address the harsh power imbalance between the land grabbers and the dispossessed (Sánchez 2017, 169).

Restitution or redistribution?

Most of the political obstacles to the implementation of a land restitution policy were eliminated when President Uribe stepped down in 2010 after two presidential terms. Policy networks built in the preceding years, particularly in the PPTP, were instrumental in advocating for a restitution agenda in the new administration. Overall, the coalition in Congress and the entire Santos administration were committed to the passage of a Victims Act and valued the political and symbolic importance of such a text. However, most governing elites were also careful to limit the agenda to the land restitution problem, wary of extending the debate to the issue of redistribution.

A 'revolutionary' law

In June 2010, Juan Manuel Santos, former defence minister and Álvaro Uribe's designated successor, won the presidential elections. While he had campaigned on a platform of continuity with his predecessor and political patron, Santos began to move away from Uribe as soon as he came to power. The contrast was clear-cut when seen from the point of view of the power networks supporting his administration. While Uribe had ruled for eight years with the support of regional barons, many of whom had close ties with paramilitary militias, Santos favoured the technocratic networks in the capital from which he himself had emerged. Senior executives from the private sector and expatriate professionals in the United States and Europe were called upon to join the administration. In just a few months, the former government's networks were ousted; the break-up was all the more brutal as Santos did nothing to halt the criminal investigations against members of Uribe's close circle for corruption and links with paramilitary groups.

Although, as a former defence minister, he claimed credit for the military successes of the Uribe era, Santos adopted a discourse of appeasement immediately after his election. This included building a large centre-right coalition in Congress, as well as finding an agreement with the leftist opposition to push a certain number of top-priority bills through the legislative agenda. One of these bills concerned reparation for victims and land restitution (Uprimny and Sánchez 2010, 307). In his August 2010 inauguration speech, Santos explicitly expressed the link between this bill and a broader agrarian and rural policy that aimed to 'pay the historical debt' accumulated by the country towards its rural poor, who had been the most affected by war (Dávila 2018, 325–7). He put the Ministry of Agriculture in the hands of Juan Camilo Restrepo, a long-time opponent of Uribe who, although a member of the Conservative Party, defended a very progressive agrarian agenda (Grajales 2019). Restrepo's roadmap included two legislative priorities: a victims and land restitution bill, and a broader rural and agrarian development bill; both texts were seen as complementary and as furthering a transformative agenda for the countryside (Restrepo and Bernal 2014).

Enacting the Victims Act only took nine months – a rather quick process considering what the previous Liberal bill had been forced to endure and the usual bargaining process in Congress. The text was endorsed by a majority of Congress members, and only the most radical rightist wing of Uribe supporters manifested vocal opposition to the bill. Most of the national press participated in the consensus surrounding the necessity and legitimacy of the project, only objecting on points related to the administrative organization or implementation of the measures (Vallejo and Montoya 2017, 156). In many ways, the text marked the victory of the mobilization in favour of transitional private law that had been going on for years inside the state. PPTP lawyers, along with the other prominent members of Bogota's progressive legal community, abundantly contributed to drafting the bill. All the legal principles featured in the Liberal bill that had been promoted during the previous years were included in Santos' version. The text went even farther, as it introduced a new principle: good faith with due diligence (*buena fe exenta de culpa*), providing that, in order to be recognized as legitimate owners of plots claimed by dispossession victims, current occupants not only needed to prove that they had acted 'within the boundaries of the law', but had to provide the proof of their 'extra care or diligence' in making sure that the property they were purchasing had not been violently grabbed from someone else (Dávila 2018, 331). To ensure that victims would be able to defend their rights in court, the law created a land restitution unit tasked with documenting cases of dispossession and bringing them before a judge. The unit would act as the victim's legal counsel, thus recognizing the necessity of putting the law and the state on the side of victims (Gutiérrez 2019).

However, while creating new rights for people affected by violent dispossession, the Victims Act also contributed to limiting the agrarian debate to the legal treatment of unlawfully acquired assets and reparations for victims of armed violence. As the government strived to build a large consensus around the law, it insisted on the fact that the restitution policy was only aimed at returning dispossessed lands to their legitimate owners (Vallejo and Montoya 2017). Various safeguards were built into the law in order to provide assurances to large landowners that it would not become a pathway to broader demands for redistribution. For instance, the law provided that if an agribusiness entrepreneur capable of proving her 'good faith with due diligence' was operating a farming project on dispossessed land, the judge would impose a leasing partnership between the peasant and the entrepreneur in order to protect the latter's investment.

The limits of restitution

In spite of these various safeguards, the early 2010s political climate was favourable to a larger debate on the diverse forms of dispossession, and the embeddedness of land grabbing within the larger political economy. As a matter of fact, one important consequence of these political transformations was to highlight that land grabbing was not only a matter of violent and outright spoliation, but also a consequence of the lack of direct state presence, widespread corruption, and weak administrative capacity. This was a slippery slope; as the first chapter showed, land accumulation and the development of agrarian capitalism are intrinsically related to the state's (at times intentional) failure to counter the interests of the powerful while (also intentionally) refraining from protecting those of the powerless. As the debate on land dispossession unfolded, the connection was made between the most egregious forms of accumulation and less extraordinary – but also illegitimate – ways of grabbing land. Controversies then focused on the accumulation of public-domain land, particularly in the region of Orinoco, Colombia's last agrarian frontier (Grajales 2020). The characteristics of this controversy and the government's vigorous reaction, eliminating the chance of any further debate, illustrate how important it was to many powerful actors within the administration to make a clear distinction between spoliation and legitimate accumulation. This chapter's final pages will be devoted to this point.

From the first months of the administration, Minister Juan Camilo Restrepo was tasked with the mission of addressing the corruption of existing land governance agencies. One of the top priorities was the reform of the Colombian Institute for Rural Development, or Incoder, arguably one of the most corrupt and inefficient institutions in the land governance sector (Restrepo 2013). Incoder had been created in 2003, in the midst of a radical reform of the agricultural sector, through the merger of Incora (the Colombian Institute for Agrarian Reform) with three agricultural development agencies. This institutional merger had been interpreted as a further step towards the state's withdrawal from rural development, and involved drastic cuts to overall budgets, staff, and territorial presence (Machado 2017, 74).

Restrepo appointed people of confidence in key positions inside the agency. Some of them, like director Miriam Villegas and deputy director Jhenifer Mojica had a civil society background, having spent most of their career working for NGOs dedicated to rural development. This was an unprecedented feature, which attracted young professionals with a similar trajectory interested in transforming the institution. In addition to this staff renewal, Incoder's management established an internal unit of criminal law attorneys dedicated to uncovering the corruption networks within the agency

Over its first months in office, this team discovered a complex network of corrupt relationships linking politicians, heads of local divisions, paramilitary leaders, and businesspeople. Many cases were found in which these various actors had conspired to legalize the violent dispossession of land belonging to forcibly displaced people. In the first months of 2012, more than 900 investigations were opened by the attorney general's office on the basis of information provided by Incoder management. The Magdalena division, where evidence of collusion between local management and paramilitary leaders was overwhelming, was one of the hot spots in this internal police operation.

Over the course of these investigations, Incoder's management and its legal team ran up against an even larger variety of techniques for unlawful accumulation, which went beyond paramilitary dispossession. One of them was the accumulation of public-domain land, made possible by the institutional weakness of the agency and the support of networks of corruption. Suddenly, public-domain land accumulation became a fashionable public problem. It embodied the state's incapacity to regulate access to land and the ways in which this had contributed to an unequal property structure. Other public agencies, such as the Notary and Registry Superintendency (SNR), in charge of the surveillance of notarial and land registry offices, also became involved in the quarrel. The SNR's director was an ambitious politician, and he tried to seize the spotlight in the fight against public land accumulation. This agitation began to produce some results. By early 2013, 60,000 hectares of unduly accumulated land had been recovered, procedures concerning 250,000 hectares had been opened, and old pending suits concerning a surface of 800,000 hectares had been resumed (Restrepo 2013, 4).

One of the reasons for this enthusiasm can be found in the timing of the public-lands case. The sudden discovery of the massive plundered patrimony of the state brought with it the promise of an unexpected source for land distribution and resonated with the old illusion of non-antagonistic agrarian reform (see Chapter 1). This corresponded to the early stages of the FARC peace talks, when a rural distributive agenda was in the making (see Chapter 5). On the surface, implementing a more transparent land management system allowing the authorities to distribute public-domain land to landless peasants would nurture a distributive policy without touching the more established and powerful interests of the landed elite (Gutiérrez and Marín 2018). It was a perfect case of distribution without redistribution. However, as Incoder management dug into some of the most glaring cases of public-domain land accumulation, they realized that they were entering into a very sensitive terrain. The case of the Orinoco plains exemplifies this.

The Orinoco plains, a savannah region spanning around 250,000 square kilometres between the Andes, the Amazonian jungle, and the Venezuelan border, has historically been perceived as Colombia's largest agrarian frontier (Rausch 2013). While the region was home to turf wars between the FARC and several paramilitary militias during the late 1990s, a security consolidation policy spurred agricultural investment over the following decade. Colombian agribusiness champions, as well as subsidiaries of large multinational corporations, settled down after 2003, making massive farmland purchases. In 2010, global giant Cargill acquired 52,000 hectares of land, while announcing a project aimed at eventually exploiting more than 90,000 hectares (Salinas Abdala 2012).

However, Incoder inquiries found most of these acquisitions to be legally flawed. As a matter of fact, most of this land was public land. While the firms had generally purchased the land from previous occupants, some sellers had no legal ownership rights, and the transactions were to be considered void. To ward off public scrutiny, most of the firms had created shell companies, each of which would purchase a certain number of plots, which prevented them from exceeding certain legal accumulation limits. The results of this official scrutiny were reinforced by other inquiries. House Representative Wilson Arias, a member of the leftist Alternative Democratic Pole (PDA) commissioned his congressional staff to carry out surveys of the investments made by agribusiness corporations in Orinoco. They identified large land acquisitions by 15 corporations; all of them had used the same shell company method.[2] Congressman Arias' accusations were further reinforced by an Oxfam report published in September 2013, which specifically targeted the role of Cargill (Oxfam International 2013).

By then, the scandal personally affected President Santos, as the inquiries disclosed the pivotal role played by one of Bogota's premier law firms, Brigard & Urrutia. They were legal counsels for various large agribusiness firms and had supposedly devised the shell companies' method. Yet, one of the firm's top associates was Carlos Urrutia, a close friend of the President and ambassador to Washington. As the scandal originated in information produced from within the administration, the Executive saw it as a political attack. Moreover, the case was politically sensitive because it hurt the image of high-profile foreign investors at precisely the moment when the government was promoting the image of a 'new Colombia', an open market with a stable business environment. By no means was the government prepared to accept for large investors who had simply played with the law to be compared to criminal mobsters who had violently grabbed hundreds of thousands of hectares of farmland. As such, the Orinoco case contributed to setting the boundaries between legitimate accumulation and spoliation. In the words of former Incoder deputy-director:

> When the scandal over the ties between Riopaila [one of the corporations involved in land-accumulation] and Ambassador Urrutia exploded, the presidency asked us for a report. We had been investigating for a long time. So we presented all the elements to the President. He told us it couldn't continue. That the priority here was investment and delivering legal security to investors. He said, 'It's one thing to prosecute Castaño, Carranza or Mancuso [paramilitary leaders]; it's another thing to attack good people'. When we [she was accompanied by Incoder's Director] left his office, we realized that we had to begin packing our bags.[3]

The shift in the political climate led to the resignation of Minister Restrepo, as well as Incoder's director and deputy director. The latter was even subjected to disciplinary proceedings. This case finely illustrates the efforts made by influential members of the administration – and the President himself – to circumscribe the land restitution agenda to the archetypal problem of violent dispossession. While the use of outright violence by armed militias, often in coordination with corrupt officials and agrarian entrepreneurs, was a problem of law enforcement and post-conflict transition, other forms of accumulation, less directly associated with armed conflict but equally constitutive of a radically unequal property regime, were to be left out of the discussion. This differentiation between spoliation and legal accumulation was based on legal considerations, but also on the promotion of an open-market economy with agribusiness investment as a primary vector of economic growth and development. The limits of the legal scrutiny on land accumulation were set by this pro-business agenda.

Conclusion

In this chapter, I have endeavoured to historically trace the emergence of land dispossession as a public problem. In the first section, I showed that, in spite of incessant flows of internal refugees pushed out of their homes by armed violence beginning in the 1980s, the Colombian state only recognized the existence of a problem of internal displacement by the middle of the 1990s. This was the result of various mobilizations, which took place in both the domestic and the international arenas. However, as important as this change of attitude might be, I also demonstrated that, until the early 2000s, the state's commitment to this problem was primarily rhetorical. While the administration was fairly successful at avoiding criticism and international scrutiny, its concrete results in the field of humanitarian assistance remained very poor.

The problem of land dispossession precisely emerged in the middle of these mobilizations in favour of an official recognition of IDPs as subjects of rights, a process I traced in the second section. This population was massively constituted of rural dwellers, many of them peasants. As such, NGOs and international development and humanitarian agencies saw the ties between forced displacement and land grabbing at a very early juncture. However, the state was much less eager to recognize this link. Bearing on important scholarly works, I traced the first attempts at creating public policies that aimed to prevent land dispossession; these attempts were the work of a discreet but very well-connected network of policy activists working within the government and enjoying the support of foreign donors. These networks, established under difficult conditions during the early 2000s, had become much more influential by the end of the decade, as the introduction of transitional justice in the Colombian legal field turned land restitution into an increasingly legitimate policy issue.

In the third section, I dealt with the ambiguous legacy of these mobilizations. While they did result in a very progressive and ambitious piece of legislation, the Victims Act, this culmination was accompanied by careful efforts to circumscribe the land problem to violent dispossession. From the onset, the Victims Act pursued the restoration of a previous state of affairs, not the overcoming of the inequality regime. As such, it was based on the conviction that it should be possible to trace a clear-cut line between spoliation and legitimate accumulation.

The next chapter will move chronologically to the most recent period, characterized by the signature of a peace agreement with the FARC guerrilla group and the very partial implementation of the resulting agrarian policies. My focus will be on redistribution. The flip side of the attention granted to dispossession and restitution was the invisibilization of a larger redistributive agenda, and so the peace talks brought with them the promise of opening up the debate. The FARC had historically been seen as a peasant guerrilla group, and the government's peacemaking team was committed to an ambitious transformative agenda. However, in the four years between the signature of the deal and the time of writing of this work, most attempts to use the agreement as a platform for transforming the land inequality regime have failed. I will now tell the story of this failure.

Notes

1 Interview with a former PPTP official, Bogota, July 2015.
2 Interview with former congressional staffer, Bogota, July 2015.
3 Interview with Jhenifer Mojica, Bogota, July 2015.

References

Allain, M. 2019. Les jeux d'échelles de l'action collective: militantisme local et solidarité internationale dans les campagnes de Colombie. *Critique internationale*, 82(1), 51–73.

Aparicio, J.R. 2009. Rumors, Residues and Governance in the Best Corner of South-America: A Grounded History of the Human Limit in Colombia. PhD diss. University of North Carolina, Chapel Hill.

Ballvé, T. 2013. Grassroots Masquerades: Development, Paramilitaries, and Land Laundering in Colombia. *Geoforum*, 50, 62–75.

Cardona-Fox, G. 2019. *Exile within Borders: A Global Look at Commitment to the International Regime to Protect Internally Displaced Persons*. Leiden; Boston: Brill.

Carranza, F., A. Gómez, C. López, and W.E. Ospina. 2019. Límites en el diseño de la restitución en Colombia. *In*: F. Gutiérrez, R. del P. Peña-Huertas, and M.M. Parada-Hernández, eds. *La tierra prometida: Balance de la política de restitución de tierras en Colombia*. Bogotá: Editorial Universidad del Rosario, pp. 39–73.

Carranza-Franco, F. 2019. *Demobilisation and Reintegration in Colombia: Building State and Citizenship*. London: Routledge.

Celis, L. 2018. *Luchas campesinas en Colombia (1970-2016): resistencias y sueños*. Bogotá: Desde Abajo.

Cohen, R. 2012. From Sovereign Responsibility to R2P. *In*: W.A. Knight and F. Egerton, eds. *The Routledge Handbook of the Responsibility to Protect*. London: Routledge, pp. 7–20.

Cohen, R. and F.M. Deng. 1998. *Masses in Flight: The Global Crisis of Internal Displacement*. Washington, DC: Brookings Institution Press.

Cristo, J.F. 2012. *La guerra por las víctimas: lo que nunca se supo de la ley*. Bogotá: Ediciones B.

Cubides, F. 2006. Proceso inconcluso, verdades a medias: para un balance de las negociaciones del gobierno Uribe con los paramilitares. *Análisis político*, 19(57), 55–64.

Daviaud, S. 2010. *L'enjeu des droits de l'homme dans le conflit colombien*. Paris: Karthala.

Dávila, J. 2018. A Land of Lawyers, Experts and "Men without Land": The Politics of Land Restitution and the Techno-Legal Production of "Dispossessed People" in Colombia. PhD diss. Harvard University, Harvard.

DNP, (Departamento Nacional de Planeación). 1995. *CONPES 2804. Programa nacional de atención integral a la población desplazada por la violencia*. Bogotá: Consejo Nacional de Política Económica y Social.

DNP, (Departamento Nacional de Planeación). 1997. *CONPES 2924. Sistema nacional de atención integral a la población desplazada por la violencia*. Bogotá: Consejo Nacional de Política Económica y Social.

DNP, (Departamento Nacional de Planeación). 1999. *CONPES 3057. Plan de acción para la prevención y atención del desplazamiento forzado*. Bogotá: Consejo Nacional de Política Económica y Social.

Elster, J. 2004. *Closing the Books: Transitional Justice in Historical Perspective*. Cambridge: Cambridge University Press.

Estrada, M. del R. and N.M. Rodríguez. 2014. La política de tierras para la población desplazada 2001-2011: de la protección a la restitución. *Estudios Socio-Jurídicos*, 16(1), 75–119.

Estrada, M. del R. and N.M. Rodríguez. 2016. Tácticas de gobierno en la política de protección de tierras de la población desplazada en Colombia: Una etnográfica del Estado. *AIBR. Revista de Antropología Iberoamericana*, 11(03), 381–404.

Garibay, D. and J.-C. Guerrero-Bernal. 2007. Identifier et interpréter une « crise extrême ». La « communauté internationale » face au conflit armé en Colombie. *In*: M. Le Pape, J. Siméant, and C. Vidal, eds. *Crises extrêmes*. Paris: La Découverte, pp. 136–50.

Gauri, V. and D.M. Brinks, eds. 2010. *Courting Social Justice: Judicial Enforcement of Social and Economic Rights in the Developing World*. Cambridge: Cambridge University Press.

Grajales, J. 2011a. The Rifle and the Title: Paramilitary Violence, Land Grab and Land Control in Colombia. *The Journal of Peasant Studies*, 38(4), 771–92.

Grajales, J. 2011b. El proceso de desmovilización de los paramilitares en Colombia: entre lo político y lo judicial. *Desafíos*, 23(2), 149–94.

Grajales, J. 2016. Quand les juges s'en mêlent. Le rôle de la justice dans la démobilisation des groupes paramilitaires en Colombie. *Critique internationale*, 70, 117–36.

Grajales, J. 2019. Les terres de la paix. Politiques de l'aide, politiques foncières et sortie de conflit en Colombie. *Gouvernement et action publique*, 8(4), 25–47.

Grajales, J. 2020. A Land Full of Opportunities? Agrarian Frontiers, Policy Narratives and the Political Economy of Peace in Colombia. *Third World Quarterly*, 47(7), 1141–60.

Gutiérrez, F. 2019. Lo bueno, lo malo y lo feo de la restitución de tierras en Colombia: una lectura política e institucional. *In*: F. Gutiérrez, R. del P. Peña-Huertas, and M.M. Parada-Hernández, eds. *La tierra prometida: Balance de la política de restitución de tierras en Colombia.* Bogotá: Editorial Universidad del Rosario, pp. 9–37.

Gutiérrez, F. and M. Marín. 2018. Tierras en el posconflicto: ¿en el fondo cuál es el problema? *Análisis político*, 31(92), 18–38.

Gutiérrez, F., R. del P. Peña-Huertas, and M.M. Parada-Hernández. 2019. *La tierra prometida: Balance de la política de restitución de tierras en Colombia*. Bogotá: Editorial Universidad del Rosario.

Hayner, P.B. 2010. *Unspeakable Truths: Transitional Justice and the Challenge of Truth Commissions.* London: Routledge.

Human Rights Watch. 2005. *Las apariencias engañan. La desmovilización de grupos paramilitares en Colombia.* New York, NY: HRW.

IACHR, (Inter-American Commission on Human Rights). 1999. *Third Report on the Human Rights Situation in Colombia.* San José de Costa Rica: Organization of American States.

Karl, R.A. 2017. Century of the Exile: Colombia's Displacement and Land Restitution in Historical Perspective, 1940s–1960s. *Canadian Journal of Latin American and Caribbean Studies / Revue canadienne des études latino-américaines et caraïbes*, 42(3), 298–319.

Leal Buitrago, F., ed. 1996. *Tras las huellas de la crisis política.* Bogotá: Tercer Mundo.

Lecombe, D. 2014. *« Nous sommes tous en faveur des victimes ». La diffusion de la justice transitionnelle en Colombie.* Paris: Institut universitaire Varenne.

Lefranc, S. 2009. La professionnalisation d'un militantisme réformateur du droit: l'invention de la justice transitionnelle. *Droit et société*, 73(3), 561–89.

Loescher, G. 2001. *The UNHCR and World Politics: A Perilous Path.* Oxford: Oxford University Press.

Machado, A. 1999. *El mercado de tierras en Colombia: una alternativa viable?* Bogotá: Tercer Mundo.

Machado, A. 2017. *El problema de la tierra: Conflicto y desarrollo en Colombia.* Bogotá: Penguin Random House.

Oxfam International. 2013. *Divide and Purchase. How Land Ownership Is Being Concentrated in Colombia.* Nairobi: Oxfam.

Rausch, J.M. 2013. *Territorial Rule in Colombia and the Transformation of the Llanos Orientales.* Gainesville, FL: University Press of Florida.

Restrepo, J.C. 2013. *Implementación de la política integral de tierras (2010-2013).* Bogotá: Ministerio de Agricultura y Desarrollo Rural.

Restrepo, J.C. and A. Bernal. 2014. *La cuestión agraria: tierra y posconflicto en Colombia.* Bogotá: Penguin Random House.

Rodríguez Garavito, C. 2014. *Law and Society in Latin America: A New Map.* London: Routledge.

Rodríguez Garavito, C. and D. Rodríguez Franco. 2010. *Cortes y cambio social. Como la Corte Constitucional transformó el desplazamiento forzado en Colombia.* Bogotá: Antropos.

Saffon, M.P. 2010. The Project of Land Restitution in Colombia: An Illustration of the Civilizing Force of Hypocrisy? *Estudios Socio-Jurídicos*, 12(2), 109–94.

Salinas Abdala, Y. 2012. El caso de Colombia. *In*: FAO, ed. *Dinámicas del mercado de la tierra en América Latina y el Caribe: concentración y extrajerización.* Rome: FAO.

Salinas Abdala, Y. and J.M. Zarama Santacruz. 2012. *Justicia y paz. Tierras y territorios en las versiones de los paramilitares.* Bogotá: Centro Nacional de Memoria Histórica.

Sánchez, N.C. 2017. *Tierra en transición: justicia transicional, restitución de tierras y política agraria en Colombia.* Bogotá: Antropos.

Tate, W. 2007. *Counting the Dead: The Culture and Politics of Human Rights Activism in Colombia.* Berkeley, CA: University of California Press.

Uprimny, R. and N.C. Sánchez. 2010. Los dilemas de la restitución de tierras en Colombia. *Estudios Socio-Jurídicos*, 12(2), 305–42.

Uprimny, R., C. Botero, M.P. Saffon, and E. Restrepo. 2006. *¿Justicia transicional sin transición? Verdad, justicia y reparación para Colombia.* Bogotá: Dejusticia.Uprimny, R. and N.C. Sánchez. 2010. Los dilemas de la restitución de tierras en Colombia. *Estudios Socio-Jurídicos*, 12(2), 305–42.

Vallejo, M. and C. Montoya. 2017. *Los cercos del debate sobre restitución de tierras: Encuadres retóricos de la Ley 1448 de 2011 en la prensa colombiana nacional y regional.* Bogotá: Editorial Pontificia Universidad Javeriana.

Vidal López, R.C. 2007. *Derecho Global y Desplazamiento Interno: Creación, usos y desaparición del desplazamiento forzado por la violencia en el Derecho Contemporáneo.* Bogotá: Editorial Pontificia Universidad Javeriana.

Weiss, T.G. and D.A. Korn. 2006. *Internal Displacement: Conceptualization and Its Consequences.* London: Routledge.

WFP, (World Food Programme). 2001. *Estudio de Caso de Las Necesidades Alimentarias de La Población Desplazada de Colombia.* Rome: World Food Programme.

5

THE UNFULFILLED PROMISES OF THE HAVANA PEACE AGREEMENT

In September 2012, the government and the FARC announced, for the first time in a decade, the official start of peace talks. When the agenda was agreed upon, rural development and agrarian policies ranked high on the list of priorities. The first of the six chapters of the agenda agreed upon in May 2013 was entitled 'Towards a New Colombian Countryside: Comprehensive Rural Reform' (CRR, often simply called the 'agrarian chapter'). It included measures for pro-poor agrarian development, access to land, and revamped state presence in marginal rural areas. The Havana talks promised to open up the agenda, linking peace not just to the fate of the victims, but more broadly to improved livelihood for the rural poor (García Trujillo 2020).

In 2014, President Juan Manuel Santos ran for re-election; harshly criticized by right-wing opponents of the peace process and facing competition from centrist and leftist parties, Santos used the peace talks as his main campaign argument. He succeeded in putting them at the centre of the agenda and was able to garner support both from urban progressive constituencies and rural voters in the regions most exposed to armed strife. As the 2014 presidential elections gave greater visibility to the peacebuilding agenda, policy professionals who had been working on the prospective implementation of post-conflict agrarian policies saw a window of opportunity. Enacting a post-conflict transition was a source of funds and visibility across a range of policy sectors. For many, post-conflict transition became much more than just the prospect of FARC demobilization. It was an opportunity for change, a promise of modernization and social transformation that seemed to take form in September 2016, when the final accord was signed (Rettberg 2020).

However, these hopes were short-lived. On 2 October 2016, the referendum intended to ratify the accord failed to win the ballot. The political capital associated with post-conflict transition suddenly evaporated. Participating in peacemaking efforts was no longer an advantage – it was a liability. Policy professionals' capacity to pursue the objectives of post-conflict agrarian policy was lost precisely at the moment when CRR was supposed to be set in motion.

In May 2018, the Colombian people went to the polls to elect Santos' successor. The elections were a clear success for the right-wing candidate and designated heir of Alvaro Uribe, former Senator Iván Duque. Duque's campaign had been built around the same network that had fought against the 2016 peace agreement, and he abundantly exploited the anti-FARC sentiments of a certain segment of voters. While he generally left the most controversial statements to his campaign's lieutenants, the promise to crush the peace agreements was clear.

Two years after Duque's election, at the time of writing of this book, it can hardly be argued that this promise has been fulfilled. As with most peace agreements around the world, the Colombian case has been characterized by partial implementation and high levels of post-accord violence. However, my priority here will not be to evaluate the current government's actions, but rather to analyse, from a broader perspective, the events that took place between the CRR agreement of 2013 and the early Duque presidency. I will argue that, while the peace talks gave momentum to the pursuit of a mildly redistributive agenda, all traces of redistribution have now been eliminated. This, I will demonstrate, is not simply the consequence of a regressive elite's campaign against redistribution. In the Colombian context, much less was necessary to dismiss such a fragile idea. In fact, the exceptional circumstances unleashed by the promises of post-conflict transition were rapidly superseded by a return to normal – i.e., the defence of a proprietarian social order.

My analysis is based on an alternative definition of post-conflict transitions. Most practitioners see post-conflict transitions as fundamentally determined by a decrease in levels of violence, often obtained through political settlements between warring parties and generally accompanied by a set of peacebuilding techniques, such as DDR programmes (disarmament, demobilization, and reinsertion), free elections, and transitional courts of justice. In addition to these core characteristics, scholarly literature mentions the presence of a broader political and economic agenda for transformation, often in the form of liberalization and deregulation, an ensemble brought together under the encompassing notion of 'liberal peace' (Ottaway 2002; Campbell *et al.* 2011; Richmond 2012; Richmond and Mac Ginty 2015). For my part, I contend here that espousing a substantialist definition of post-conflict transitions can be misleading. This is particularly true when researchers focus less on the political dynamics of armed de-escalation than on the host of peace policies and policy professionals gravitating around the very idea of transition (Pugh *et al.* 2008; Suhrke and Berdal 2013; Berdal and Zaum 2017). Most of these policies cannot be simply described as reactions to the problems posed by armed violence, but should better be understood as complex constructions that combine existing solutions, (sometimes) new ideas, and (often) new justifications (Grajales 2019; Grajales and Jouhanneau 2019).

This leads me to adopt two complementary approaches. First, I consider that we should be concerned by the mechanisms behind the redefinition of public policies, as certain sectors of public action are associated with post-conflict transition.

As such, post-conflict periods should be seen as a 'framing' (Schön and Rein 1995; Rein and Schön 1996; Surel 2000) that offers a redefinition of policy problems and opens up access to greater financial, political, and bureaucratic resources. These redefinitions have multiple effects, such as transformations in the actors who appear to be legitimate 'owners' of public problems (Gusfield 1989). A second, complementary approach unfolds from this idea, as post-conflict transitions can also be defined as configurations, in the sense of Norbert Elias (2001). Under the abstract and rather misleading idea of 'transition', more concrete social realities can be observed, such as new divisions of labour within institutions and new distributions of resources between actors and policy sectors, as well as new political and administrative hierarchies, all legitimized in the name of peace.

This chapter therefore aims to trace the different forms of association between post-conflict transition and land. The first section describes how land was made an integral part of the peacemaking agenda and the effects of this association on the attention granted by influential policy actors to previously marginal agrarian problems. The second section will inquire into the reasons for the rapid closure of this window of opportunity, linking it to two key moments: the peace agreement referendum and Iván Duque's presidential victory. Lastly, the third section will explore the policymaking mechanisms at work, showing how problems of land inequality and redistribution were swiftly eliminated from the outcomes of the Havana peace talks.

A time for change

How are some policy sectors defined as a part of the negotiated path out of war? How do specific policies become instruments of peacebuilding? Of course, in some cases, such as with DDR or transitional justice, linking issues in this way might seem natural. But when we move to the margins of post-conflict transitions, to policy domains that are less directly determined by the dynamics of armed violence, this link becomes tenuous. Looking at this 'ordinary aftermath' of armed conflicts requires a more acute examination of how policy issues are constructed as a part of the peacebuilding package (Grajales and Jouhanneau 2019). This section investigates how land became an important part of the peace agreement, the effects of this relationship on the capacity of rural development professionals to promote a policy agenda, and the role of foreign development agencies in this dynamic.

Bringing land to Havana

The land issue was brought to the Havana talks through three main channels. The first is the historic connection that has existed between peacemaking and land, especially in the case of the FARC, which has often been defined as a 'peasant guerrilla'. Beginning with the 1964 'agrarian programme', the FARC has consistently defended an agrarianist agenda composed of both demands for state intervention in favour of impoverished peasants and limitations on land accumulation and land purchase by foreign investors (for a historical account, see García Trujillo 2020, 61). In the same vein, peace negotiations with the FARC had always integrated an agrarian component throughout the 1980s and 1990s (Restrepo and Bernal 2014, 180-sq).

However, the representation of the FARC as a 'peasant guerrilla' group (Richani 2002; Leech 2011) is not fully accurate. True, the FARC rank-and-file was massively composed of people from poor agrarian backgrounds, very often settlers from the agrarian frontiers where the state has historically been either absent or brutal in its treatment of the populace (Aguilera Peña 2014; Carranza-Franco 2019). However, for most of its history, the FARC peasant agenda has been detached from the political debates that have animated peasant organizations. The FARC were marginal to the massive peasant mobilizations of the 1960s and 1970s, and were largely sidestepped by the peasant protests of the 1990s (Zamosc 1986; Ramírez 2011; Celis 2018). By the 2000s, the Colombian peasant movement's claims had become structured by its efforts to shield itself from the dynamics of war and its adoption of an agenda centred around the issues of autonomy, the environment, and territory, largely inspired by international debates on food sovereignty, agro-ecology, peasant identity, and territorial rights (Allain 2016; Roa-Clavijo 2021). In Havana, FARC negotiators' initial claims followed the classic attacks against '*latifundia*' and foreign (especially US) investment (Bermúdes 2018, 83). The peasant movement agenda only began to be integrated into the FARC's discourse during the peace talks.

The second channel connecting land, violence, and rural development concerns the ways in which the Santos administration framed rural issues in its early years (for a detailed account, see García Trujillo 2020, 121-sq). Much like the Liberal reformists of the 1930s and 1960s, many in the progressive policy establishment saw the negotiations as a favourable conjuncture for putting an end to the archaic inequalities of the Colombian countryside and laying the groundwork for a more efficient and socially responsible property structure. With or without a peace settlement, these transformations were urgently needed for the nation's stability and economic development (Bermúdes 2019, 33–5). These and other ideas about land distribution, rural development, and support to peasant farming were defended in Cuba by widely recognized experts, most notably sociologist Alejandro Reyes, who was called to Havana as a special advisor to assist with negotiations on agrarian issues (Reyes 2013).

The third channel through which the issue of land was brought to the Havana peace talks was embodied by key members of the negotiation team, who each shared the conviction that rural poverty and landlessness were the primary underlying cause of armed conflict. This resulted from their experience as policymakers and their professional trajectories within the state. This is exemplified by the trajectories of two men: Sergio Jaramillo and Álvaro Balcázar. Jaramillo became the government's High Commissioner for Peace in September 2012. He had served as Deputy Minister of Defence tasked with human rights and international affairs when Juan Manuel Santos was at the head of the Ministry under the second Uribe administration. One of his most relevant achievements was the development of a 'consolidation' policy implemented by Álvaro Balcázar, a development economist and one of Jaramillo's closer collaborators within the ministry. With the support of USAID's Office for Transition Initiatives (OTI), the programme focused on bringing coordinated state action to zones recently occupied by the army, with measures such as social benefits, rural infrastructure building, land titling, and government-backed loans. The guiding idea was that state presence and social legitimacy would 'cut off the guerrilla group's oxygen'.[1] In 2012, Balcázar joined the negotiation team in Havana to provide support to the agrarian agreement.

These three channels of connection explain that, from the outset of negotiations, the agrarian chapter was identified as instrumental to putting the peace talks on track. For members of the government, the topic was meant to become a 'golden bridge' enabling the two parties to find fruitful compromises and create mutual trust (Bermúdes 2018, 2019; García Trujillo 2020). The government was comfortable with elaborating an agenda for rural development, as most of these policies were being defended from within the progressive establishment. As for the FARC, the agrarian chapter enabled them to show quick results and to claim the reformist agenda as their own. While the CRR policy was well behind the historic demands of the guerrilla and had a very mild redistributive component, FARC negotiators understood that these policies went beyond the commitments made by any recent government towards the peasantry. They were also realistic, as they knew that the first point was an 'easy win', compared to the much more difficult discussions that lay ahead on intractable issues such as demobilization and transitional justice. Lastly, the FARC felt that their own agrarian agenda did not belong to the past, but could become a key part of their post-conflict political platform (for a much more detailed account, see García Trujillo 2020, 113-sq). The agrarian agreement was then seen by FARC negotiators as providing a legal and institutional framework that would in the future enable them to push a more radical agenda when they had become a political party (Reyes 2013).

A window of opportunity

As the outcome of these historical considerations, the agrarian chapter did not introduce a radical rupture in policymaking. Table 5.1 summarizes the composition of the CRR policy.

This summary reflects what one of my interviewees, an executive officer at DNP in charge of rural development, told me in October 2015: 'When I read the agreement, my first reaction was – this basically tells us that we must do what any modern state should do'.[2] As a matter of fact, contrary to what its critics might claim, the agreement was not meant as a threat the proprietarian social order. Rather, its intent was to approach issues such as land concentration and landlessness indirectly, for instance, by increasing land taxation and bolstering small peasants' purchasing power. In practice, the CRR method generally consisted of taking existing policies and scaling them up, making them more binding or allocating additional resources to their implementation. Consequently, framing a policy as a part of the

TABLE 5.1 The comprehensive rural reform

Thematic area	*Provisions*
Land access and use. Idle lands. Land formalization. Agrarian frontier and protected areas	Land fund Other land access mechanisms: subsidies, special loans, usage rights Land formalization Land dispute resolution. Creation of agrarian courts. Strengthening food production Modernization of the rural cadastre Closure of the agrarian frontier, territorial planning, and environmental protection Peasant reserve zones
Territorial development programmes (PDET)	Territorial development programmes
National plans for CRR	Road infrastructure: development of tertiary roads Irrigation infrastructure Electricity and internet access in rural areas Rural health Rural education Housing and access to safe drinking water Support for agrarian production. Solidarity and cooperative economy. Technical assistance, credit, marketing support Rural labour: formalization and social security The right to food

peacebuilding package eventually became a way of obtaining financial and political resources. This framing mechanism created a window of opportunity, as policy-making professionals and policy agencies perceived peace as a useful banner.

At the same time, there were limits to this scaling-up dynamic. For instance, in late 2013, the National Statistics Department (DANE) was ready to launch a rural census, the first since 1970. Providing data on the living conditions and the livelihoods of rural populations was critical to achieving the objectives of the peace agreement. However, the project was suddenly dropped as the Treasury withdrew the funds that had already been appropriated, leading to the resignation of DANE's director as a sign of protest. Such an episode illustrates that during the early stages of the agrarian agreement's implementation, the dominant attitude was to 'wait and see' (García Trujillo 2020, 186-sq). The agrarian chapter was only the first out of six intractable issues including transitional justice and drug policy. The government itself had insisted on the rule that 'nothing was agreed until everything was agreed' (Bermúdes 2018; García Trujillo 2020). As a result, in the aftermath of the 2013 agrarian agreement, little changed in Bogota.

The campaign for Juan Manuel Santos' re-election changed the game, giving greater political visibility to the issue of negotiations. Santos decided to make peacebuilding the central issue of his platform, thus giving greater visibility to the

Havana talks. This move also resulted greater political capital for those involved in post-conflict policymaking. Then, as Santos was re-elected, the idea that the agreement was a close perspective became concrete, strengthening the preparatory work in Bogota.

This resulted into the formation of a policy network made up of officials from various agencies and ministries who would push for the Havana commitments to be translated into public policies. This network operated at two levels. At a decision-making level, the government designated deputy ministers in each ministry concerned by the implementation. These officials would be in charge of kick-starting the process, and they would be directly accountable to the President. But much more importantly for the structuration of the public policy, the network brought together – at a technical level – highly trained officials, often young professionals, from within various ministries and agencies. The task of coordinating this network fell upon the staff of the High Commissioner for Peace's office, who were also members of Havana's negotiating team. The first step in their policy action was to extract commitments and budgetary appropriations from the agreement and bring them together in a 'framework for implementation' (*plan marco de implementación*).

An additional step in the materialization of the agrarian agreement's commitments was taken in December 2015 with the creation of three executive agencies in charge of CRR: the National Lands Agency (ANT), the Rural Development Agency (ADR), and the Territorial Rehabilitation Agency (ART). The ANT would steer the implementation of policies linked to land ownership, the ADR would be in charge of supporting the development of agriculture, with a special focus on family farming and vulnerable households, and the ART would be dedicated to scaling up state efforts in the most vulnerable areas of the country. The creation of the three agencies was directly overseen by the President's Cabinet Chief, María Lorena Gutiérrez (García Trujillo 2020, 192). All the technical aspects were managed from the rural development division of the DNP. This was a meaningful sign for close observers of Colombian administration, as the DNP's bureaucratic power and technocratic prestige largely outshines that of the Ministry of Agriculture.

The creation of these agencies was an important step. It fuelled the negotiations and sent the message to the FARC that the government was motivated to move forward with the implementation phase. At the same time, it created new bureaucratic spaces within the state filled with people committed to moving forward with the agrarian agenda. Ultimately, it was a way of circumventing the administrative inertia of existing institutions and established policy sectors.

International support

The perception of a changing conjuncture linking land, rural poverty, and peacebuilding drew much interest from the donor community. Until 2011, most official assistance agencies' attention was turned to humanitarian action, primarily with a focus on assistance to IDPs and other victims of violence. This bias was reinforced by the attitude of the Uribe administration, which pursued the twofold objective of extracting funds from foreign donors and limiting their influence with regard to major policy orientations. This was generally accomplished by confining them to subjects seen as 'technical' and 'non-political' (McGee and Heredia 2012). However, there were exceptions to this rule, as some foreign agencies were active in pushing for structural reforms. The best example of this attitude is provided by the United Nations Development Programme (UNDP), which since early 2003 had consistently advocated for a comprehensive policy in favour of impoverished rural areas. UNDP's human development report for Colombia, released in 2011 and entitled 'Rural Colombia: Reasons for Hope' made a very explicit link between land inequality, rural poverty, and violence (UNDP 2011). The report's release illustrated the change in the political climate. While former President Uribe had violently criticized foreign agencies' attempts to participate in policy discussions on sensitive areas, President Santos attended the inaugural conference and praised the UNDP's and other donors' contribution to peacebuilding efforts in the Colombian countryside.[3]

In the early months of his administration, President Santos expressed his will to confer a more central position to foreign donors in the peace process. Improvements in the area of security, particularly once the Havana talks had begun, along with various signs of the government's increased commitment to human rights, fed a transitional momentum. According to a discourse endorsed by most donors from 2012, Colombia was on the right track to transition from war to peace (Grajales 2019). This entailed a transformation of foreign aid, which it was urged should be less focused on emergency reaction and much more dedicated to longer-term development issues. The signature of the agrarian chapter was an important milestone in this process. Foreign agencies had finally been called upon to directly contribute to designing the policies that would bring peace to the country's rural areas. For development professionals, some of whom had supported grassroots peacebuilding projects in very hostile national and local contexts (Castañeda 2014), this was a welcome recognition of their efforts.

In July 2015, I was in Bogota for a research trip. When I met with the staff of the most active donors, the recurrent discourse from those I spoke with was that foreign assistance was going through a twofold transformation.[4] On the one hand, they evoked a transition, at least in terms of programme design, from humanitarian aid to economic development. On the other hand, most of these foreign aid professionals were seeking to redefine their position as joint creators of public policies, which they believed was made possible by the fact that their role was now less characterized by humanitarian emergencies and human rights advocacy. This was fully compatible with the vision of economic emergence being touted by the Santos government, as the country requested its admission to the OECD and the administration showcased the numerous investment opportunities fostered by post-conflict transition (Grajales 2020a). In the words of a European aid programme manager:

> In the long run, aid policy will disappear in favour of work on more equitable trade relations. This is reflected by the shift in our delegation's administrative organization: initially there were three units [on conflict and humanitarian assistance issues], whereas now there is only one unit with seven staff members. In the long run, we will all have to migrate to commercial policy units. This is primarily the result of the development of Colombia's trade potential.[5]

Recent developments in European cooperation programmes are a good illustration of these transformations. The 2002–2007 multiyear programme was entirely focused on grassroots development. Its main partners were NGOs and organizations involved in social movements. The EU's flagship programme consisted of 'peace laboratories', policy initiatives that aimed to protect and empower impoverished and vulnerable communities (Castañeda 2014). Since the middle of the decade, the stabilization of the security situation and the negotiation of a trade agreement between the EU and Colombia have bolstered the convergence between development aid and the economic agenda. While the 2008–2013 plan maintained the peace laboratories programme, it created closer ties with state institutions. By then, Colombia was becoming a global advocate in favour of autonomy and aid 'ownership' for recipient countries (McGee and Heredia 2012). In line with this transformation, but also reflecting the growing belief in the country's 'transition' out of war, the 2014–2017 plan introduced budget support programmes, thus aligning international aid with the government's policy priorities. Cooperation funds were therefore no longer intended to fund independent initiatives, but to support public expenditures. Interviewed EU programme managers insisted on the link between the improved security situation and the shift in the approach used to provide assistance. By 2015, the recurrent discourse was that Colombia was no longer a 'sick country' in need of urgent assistance, but a soaring nation whose process of economic emergence and peacemaking should be supported.

As we have seen, both from the point of view of domestic bureaucracies and from that of foreign donors, the representation of Colombia as a promising example of post-conflict transition was becoming widespread by the mid-2010s. After President Santos' re-election in 2014, there was widespread belief that the Havana talks would result in a positive outcome, and a solid 'coalition for peace' seemed to support the negotiations. This had a substantial impact on policy professionals, as associating policy sectors and instruments with the peace effort was becoming a guarantee of renewed funding and bold political support.

Closing the window of opportunity

However, these promising auspices failed to bring about lasting change. Santos' second term fell well short of objectives in terms of structural reforms. The failure of the peace agreement referendum took a heavy toll on the government's political capital in pursuing the implementation of the peace accords. Lastly, the politicization of the Havana agreement bountifully served the interests of the radical right, propelling the campaign of the regressive Centro Democrático, the party of former President Uribe. The following pages show how the implementation of CRR was ultimately hindered by this political climate.

Lost momentum

While Santos' re-election campaign contributed to placing the peace agreement at the centre of the political agenda, the circumstances of his victory would prove highly detrimental to the government's capacity to implement change (García Trujillo 2020). In the first round of the presidential elections, Santos won a mere 23 per cent of the vote, finishing second behind the candidate for the Centro Democrático. His final victory was the result of support from centre and leftist groups, whose main reason for supporting Santos was the peace agreement, but it also required hard bargaining with the rightist Cambio Radical, and with sectors of the Conservative Party, whose approach to the peace talks was opportunistic at best. In order to rally their support and ensure that they did not adopt the same anti-Havana stance that had been so successful for Uribe, Santos had to abundantly distribute chunks of bureaucratic power. Months before the election, Santos handed over key administrative positions within the agrarian sector to Conservative Party politicians. Institutions that would become instrumental to the implementation of CRR, such as Incoder or the Agrarian Bank, ended up in the hands of seasoned Conservative electoral barons, who were primarily interested in bolstering their capacity to distribute jobs among their clients (Bermúdez 2013; García Trujillo 2020).

The conditions of the re-election weakened the governing coalition. In the aftermath of Santos' second inauguration, the most opportunistic members of the majority – the Conservative Party and Cambio Radical – began to prepare Santos's succession. Meanwhile, the Centro Democrático had floundered in the presidential race, emerging as the most virulent opponent of the peace talks and the leader of the reactionary right. Paradoxically, while the politicization of the Havana negotiations shored up the administration's capacity to push post-conflict policies in the political agenda, it also made the agreements more politically fragile, polarizing the political spectrum and exposing them to partisan attacks. Contrary to past instances of peace negotiations in recent Colombian history (Chernick 2008), there would never be a semblance of political union behind the Havana talks (Gutiérrez 2019). From the moment they became a matter of political debate, they fell victim to the various cleavages structuring the partisan arena.

The politicization of the peace agreement and the overarching consequences of their political instrumentalization would become glaring as the final agreements were brought forward to be ratified by referendum (Rettberg 2020). The Centro Democrático and some prominent Conservative figures led the 'No' campaign. Members of the majority refused to openly support the 'Yes' campaign, knowing that some of their electorate was highly sensitive to Uribe's anti-FARC rhetoric. While the 'Yes' campaign was probably overconfident and failed to generate massive mobilization for the agreement – participation stalled at a mere 37.43 per cent – the opposition was successful in tying the referendum to other political questions and turning the vote into a campaign against both Santos government and political liberalism at large (Botero 2017). Radical Catholic and Pentecostal networks were mobilized against the mention of a 'gender approach' in the agreements (Beltrán and Creely 2018). In addition, the possibility of distributional policies in favour of former combatants was instrumentalized by the leaders of the 'No' campaign to fuel social rivalries (Basset 2018). As such, they succeeded in mobilizing a working-class electorate that feared that it might be stripped of its meagre social benefits, which would be redirected – or so campaign spots claimed – to demobilized guerrillas. Agrarian issues played an important role in the 'No' campaign. Agribusiness companies were among its main donors (Verdad Abierta 2016), and emblematic policies of Santos' liberal turn, such as land restitution, were violently attacked and falsely associated with the Havana agreement. This can be illustrated by Álvaro Uribe's numerous statements underscoring what, for him, were the threats posed by the agreement to the sanctity of private property. Days before the referendum, Uribe would publish a message to his several million Twitter followers claiming that: 'If you vote Yes on October 2, you are approving that the Santos government's expropriation of land from people who have acquired it in good faith. If you vote No, you can avoid it' (Vallejo and Montoya 2017, 174).

After the referendum's failure, the government found itself in an intractable situation. While the FARC's leadership renewed its commitment to the agreement, the text had to be modified to include some of the opposition's demands. This was a challenging task, as the 'No' campaign had articulated extremely wide-ranging demands, often resorting to false pretences that were nowhere to be found in the text. Even more consequential was the risk of empowering the leaders of the 'No' camp, turning them into the representatives of an electorate whose only common denominator was the vote they had cast (Botero 2017; García Trujillo 2020). Moreover, the situation was a win–win for the agreement's opponents: not only did they succeed in exerting an influence on the accord's content – for instance, by introducing further safeguards against any limitations on private property – they were also simultaneously in a position to claim that the final agreement had ignored their claims. After several weeks of negotiations conducted in parallel with the opposition in Bogota and FARC leaders in Havana, the government announced that a second version of the text would be put to a congressional vote. Realistically, a new referendum was both too risky and impossible to organize on such short notice. However, approving a new agreement in Congress despite the first version having been rejected by the popular vote ultimately cast a shadow of illegitimacy over the final peace accord.

From the government's perspective, the peace process had come to a brutal standstill. The policy network that had been formed around post-conflict agrarian policies saw much of its political capital suddenly evaporate into thin air. In the words of the former coordinator of the working group on the implementation of CRR:

> The day after the referendum, everyone stopped. People in the ministries simply focused on other things. No one had imagined that we could lose. There was a very exaggerated confidence that we would win. Things had never been easy for the people inside the ministries who had fought to make these issues a priority. But then it became impossible. People lost interest and no one would listen to us anymore. Those who still wanted to work were powerless. Their bosses simply told them to focus on something else. So at the end, if you needed to meet with the people from the DNP, or the Ministry of Agriculture, you couldn't even manage to get anyone on the phone.[6]

A similar outcome affected the discussions in Congress. With the 2018 general elections already looming, Senators and House Representatives feared alienating a part of their electorate by seeming overly supportive of the peace agreement (Rodríguez-Ruíz 2019). Overall, both for civil servants and for elected officials, associating post-conflict policies and peacebuilding was no longer an asset; it had become a liability. This occurred at the worst possible moment, as a great deal of political support was needed to translate the agreements into concrete legislation and budgetary appropriations.

Although most of the political momentum from the Havana talks was lost, the FARC were in fact beginning to demobilize. The government's attention in the first half of 2017 was absorbed by that specific endeavour. In the meantime, all other tasks were at a standstill. Paradoxically, the supporters of the peace agreement lost even more of their meagre political capital after the end of the demobilization process; the feeling of urgency was eliminated by the fact that the guerrilla had relinquished their weapons.

By the end of 2017, there was practically no one left to defend the peace agreements. The government was exhausted after seven years in power and further worn down by a series of end-of-term corruption scandals paired with a flagging economy. The Liberal Party, which had spearheaded important legislative milestones such as the 2011 Victims Act, was consumed by the upcoming elections. Congressional opposition to the agreements was taking form, with politicians from the Centro Democrático, the Conservative Party, and other right-wing elements competing to reclaim the fruits of their anti-FARC virulence. As for the former rebels, they have been unable to rally a constituency in favour of the peace accords. In August 2017, the guerrilla group's former combatants created a political party with the same acronym that had been used for the armed movement: the Common Alternative Revolutionary Force (the FARC Party). However, the FARC leadership was consistently unable to forge ties with various sectors of the rural social movement (peasant organizations, Indigenous, and Black movements, etc.) (Acuña and Pérez-Guevara 2019). At the time of writing, research is still very scarce on the reasons for this situation. One possible explanation, however, might be that the FARC Party's failure to rally a large front of popular organizations is both a matter of past history and present conjuncture. In the past, FARC leadership often attempted to instrumentalize unarmed opposition, but it never considered social movements as powerful drivers for change. In the areas under their control, the guerrilla group whipped up support from grassroots organizations by instrumentalizing community politics for their own strategic purposes. In its post-demobilization phase, those in the top ranks of the FARC have been fully concentrated on defending their personal legal status and the agreement's provisions guaranteeing their right to political participation.

The FARC's leadership paid the price for this disconnection in the ballot box. Unlike other cases of peace agreements (including past agreements in Colombia), where the demobilized guerrilla group retained a certain degree of political influence (Garibay 2005), the FARC Party's showing in the 2018 Congressional elections was woefully inadequate, with a mere 53,000 votes cast for the FARC, representing 0.34 per cent of the ballot. The party failed to become a political force even in the places where the guerrilla group had been the main political authority for decades (Acuña and Pérez-Guevara 2019). While the agreements guarantee them five seats in both the Senate and the House for three congressional periods, their voice in Congress is marginalized by their extremely poor electoral performance.

Change of government

By the end of the Santos administration, the future seemed dim for the implementation of the peace agreement. I passed through Bogota on my way back from field research in Magdalena in August 2018. This was a couple of days after Iván Duque's inauguration. As the new president delivered a rather moderate speech, the Senate Chair, a radical right-wing politician, delivered a violent diatribe against the Santos administration, claiming that Congress was bound to modify the peace agreements in order to restore the rule of law. My interviewees were terrified by the speech, fearing that it was a harbinger of radical backlash against all they had worked to achieve.

One year later, I returned to the city to meet with a number of people from the small agrarian policy sphere. Many were rather surprised by the relative moderation shown by President Duque, at least on the agrarian agenda. While members of his party had promised to 'crush the Havana agreement', Duque adopted a rather moderate stance, claiming that he simply wanted to rectify the accord without effacing previous achievements. Moreover, the priority for the rightist coalition seemed to be the most visible and controversial aspects of the agreement, such as FARC commanders' legal benefits. Overall, agrarian policies seemed to have been downscaled in the political agenda, for better or worse; there was clearly no political will to push things forward, but neither was there necessarily the political animosity required to go about fully dismantling the meagre achievements of the previous administration. In reality, this attitude was largely sufficient to extinguish any hope of transformation.

One of the new government's first steps in downscaling CRR was to limit the ability of international donors to influence the agrarian agenda. A number of donors, such as the European Union, Scandinavian countries, and the UNDP, were fervently committed to the implementation of the most distributional aspects of CRR. Some members of the administration adopted a hostile stance towards these international agencies, which were seen as too close to human rights and environmentalist NGOs and social movements. When an audit of the ANT's operations was conducted by Duque's new appointee, the new director considered the possibility of cancelling all partnerships with the FAO and the UNDP (who operated as contractors for the agency's operations) and replacing them with private contractors.[7] The attitude was less frontal with the EU and bilateral agencies, but the government nevertheless tried to make the issue of land go away. In the words of an EU programme manager:

> Look, I do multilateral cooperation. Which means that no matter how much we push on an issue, if the government doesn't want to give it any importance, things won't happen. And that's the case with land. They don't give it the importance it deserves. Or no, even worse, they explicitly try to downplay the problem. A good example of this is that in the meetings we hold with donors involved in the agrarian agenda, the ANT director never attends. She sends one of her deputies, but she never comes personally, whereas we, in the EU, we always send our chief of operations.[8]

Limiting foreign donors' capacity to influence agrarian policy implementation often involves shaping how foreign assistance programmes' success will be measured. This means eliminating any indicators that do not correspond to the administration's priorities. This is well illustrated by EU aid. In 2019, the government was negotiating an 18 million euro assistance package with the EU, which included support for CRR implementation. The EU initially included the same indicators as the ones used for the previous assistance programme (covering the 2016–2020 period). Three indicators were related to land policies: the number of regularized titles, the number of peasant families benefitting from a title deed, and the number of Indigenous families covered by a collective title. The administration refused to include the latter two, as these can – in some cases – have a distributive dimension.

There has been a similar will to downsize agrarian issues on bureaucratically. The 2020 budget for the agencies involved in implementing CRR is a good indicator of this. The ANT, ART, and ADR each respectively suffered cuts of 18.9, 10.4, and 12.9 per cent compared to the previous fiscal year (Rodríguez Lach 2019). The Ministry of Agriculture, which controls budget appropriations for the entire agrarian policy sector, intentionally slows the implementation of the most basic aspects of CRR. For instance, one of the central pieces of the reform was the creation of agrarian courts tasked with settling a large variety of land conflicts. While this legal reform was supposed to be jointly undertaken by the Ministries of Agriculture and Justice, the former never got involved in the drafting process and the latter waited over a year before opening discussions.

The downscaling of CRR can also be analysed from the point of view of the people who are supposed to oversee its implementation. Overall, by dismissing the most qualified professionals within the upper ranks of management and appointing inexperienced officials to important positions, the Duque administration essentially dismantled the agrarian policy network. This can be observed at various levels of responsibility. At the top management level, low-profile directors with no experience on agrarian issues were appointed at the head of the ANT and ADR. In August 2018, when President Duque took office, many of my contacts inside the administration feared that he would appoint fervent opponents of CRR. Rumours spread indicating that the head of the cattle ranchers' association – who was a radical right-wing figure, one of the most prominent opponents of land restitution policy, and a man accused of having entertained links with paramilitary militias – might become the ANT's general director. Instead, the government appointed an unknown business executive, former head of the corporate social responsibility division at the palm growers association. She replaced Miguel Samper, the Santos-era director, who, although lacking any specific experience on agrarian issues, had far more political relevance. He was an important figure of the Liberal Party and the son of a former President. Within the ranks of the ANT's middle management, most Santos-era officials were dismissed. Many had started their careers in the Land Restitution Unit, a professional pedigree that was turned into a stigma by the new administration. In some cases, important positions remained vacant for months, and, with much of the operational staff also being replaced, most of agency's memory was lost. Many of those who did not lose their jobs were demoted to less influential positions. This is the case of one of my long-time sources, whom I met in 2017 when he was serving as a special advisor to the agency's director after having occupied management positions at Incoder. In 2018, he was accused by his new bosses of being a FARC sympathizer, an accusation that made him worry not only for his job, but also for his life. When I met with him again in November 2019, he was working as rank-and-file lawyer in one of the agency's divisions.[9]

This has generated a human resources issue. The ANT has continually had problems filling both management and operational positions. One year after the arrival of the new administration, five of the 20 positions within the ANT's management team were still held by interim directors. In some cases, such as the appointment of the executive director, the nomination of people who lacked any consistent experience in agrarian issues can be interpreted as a political act, a sign of the dim view the government takes of the agency and an attempt to further undermine its capacity to weigh on negotiations with other institutions. But in many other cases, the agency's HR policy is not a well-thought strategy, but simply a consequence of the political climate. If people with very poor experience on agrarian matters are appointed at various levels of responsibility, it is because most of the well-trained and highly skilled professionals on the market bear the stigma of a past engagement with land restitution or CRR. They know that their professional perspectives will be much brighter working with foreign assistance donors, the private sector, or even NGOs, rather than trying to make their way within a demonetized public agency.

Overall, in the seven years that have passed between the signature of the agrarian chapter and the time of writing, CRR has become an ordinary policy, utterly deprived of the cloak of prestige and the positive outlook that its post-conflict clout once afforded it. The new institutions and legal tools created by CRR will likely become an ordinary component of the agrarian policy sector. However, as the 'peace momentum' no longer exists, backing for the most ambitious – and sometimes redistributive – aspects of CRR has been lost. The next section dives into the details of policy design to trace the demise of redistribution.

The demise of redistribution

In addition to scaling down the entire agrarian peacebuilding agenda, the government adopted a strategy of selective implementation, primarily aimed at eliminating the redistributive – or potentially redistributive – aspects of CRR, in order to focus on technical mechanisms that might be compatible with a proprietarian social order. The reshaping of two of the CRR policy's important components provides a meaningful illustration of this: the land fund and the cadastral registry.

The land fund: from a reservoir for distribution to a transit point for formalization

One of the FARC's priorities during the negotiation of the agrarian agreement in Havana was to obtain a commitment from the government to enact a land redistribution policy. However, government representatives and guerrilla negotiators differed on the appropriate means to be used. As accounted by Bermúdes (2018, 83):

> The government recognized that the state has a constitutional responsibility to progressively provide access to land to those who need it most. However, the FARC proposed the expropriation of what they called 'unproductive *latifundia*' through mechanisms that went beyond what the existing legal framework offered, such as expropriation without compensation or tasking local bodies that they called 'agrarian reform councils' with the execution of the policy.

In response to this, government representatives insisted that any redistributive policy would have to comply with existing legislation and the Constitution, both of which protect private property from expropriation. The response to this challenge was to create a land fund that would centralize the management of land available to be distributed within the limits of the law. The quantitative objectives of the fund were a matter of bargaining between the FARC and the government. Initially, guerrilla representatives set a target of 20 million hectares, considered as unrealistic by government negotiators. The FARC reduced this demand to 10 million hectares, but finally settled on a target of 3 million hectares, after a Los Andes University study commissioned by the Ministry of Agriculture concluded that this was a reasonable objective.

The fund remained ambiguous about its transformative ambitions. For the FARC, it was a redistributive tool. Their negotiators strived to include the notion of 'democratization of access to land' and the objective of 'deconcentrating and promoting equitable distribution of land' in the final accord. The government, on the other hand, viewed the fund as distributive but not necessarily redistributive. From the outset, government officials thought that the fund could distribute land without impacting the sanctity of private property (Gutiérrez and Marín 2018; Gutiérrez 2019). The agreement provides that the fund should be filled with public domain land, land purchased by the state, land expropriated from criminal offenders, and land that has been unduly accumulated; this latter category essentially comprises the fraudulent occupation of public land by private owners. A truly redistributive measure would have been real and aggressive implementation of the expropriation of idle estates, a procedure that has existed since 1936 and was relaunched in 1968, but has been very seldom applied since the conservative turn on agrarian policies in the 1970s (see Chapter 1). However, this option was dismissed.

Government representatives had been consistent throughout peace talks in affirming that land redistribution was not an aim *per se*, and that the agreement had to be compatible with an agro-export economy essentially based on the corporate plantation model. Any form of redistribution would not be the outcome of 'democratization of land access', but simply the consequence of a fiscal policy that would discourage unproductive accumulation and encourage productive estates, irrespective of their size. The modernization of the taxation system, itself dependent on the creation of an efficient cadastral registry, was intended to suppress market distortions and reduce the price of land. In other words, redistribution would stem from a well-functioning market, not from direct intervention by the state.

Such diverse understandings of what the land fund should comprise were due to the establishment of an 'ambiguous agreement', i.e., support from all parties, which was obtained not through a consensus on the policy's content, but thanks to the ambiguity of the measure (Palier 2005). Once CRR as a whole had been scaled down to smaller proportions, this ambiguity enabled a transformation of the land fund that did away with any redistributive ambition entirely. Ultimately, the administrative practices in place at the ANT reduced the fund to a mere 'transit point' for untitled land waiting to be formalized. In the words of a former ANT senior executive:

> The land fund is a legal fiction. You always formalize land in favour of someone who is already there. This is not land available for distribution. In that sense, the fund is not a redistributive instrument, it is simply a transit point. The idea that the fund is going to be filled up with land taken from large landlords is a fantasy.[10]

The hypothesis of a rapid transition of the fund from a reservoir for distribution to a transit point for formalization is confirmed by the latest figures on the agency's action. From 2017 to 2020, the agency formalized 1.9 million hectares but allocated a mere 8230 hectares (PGN 2020, 44). As such, while a great deal of effort has been expended to achieve quantitative objectives, the redistributive dimension of the policies has been disregarded. Overall, the fund, which in the spirit of the peace agreement was supposed to allow for redistribution, became no more than a legal tool to measure the advancement of land formalization policy. The case of the rural cadastre offers a similar example of how redistributive aspects have been left by the wayside.

The cadastre: a representation of property rights?

In Colombia, the rural cadastre suffers from the kind of shortcomings common in the Global South, such as a lack of efficient updating procedures, poor management, and a lack of communication with other real estate registries. Instead of interpreting this as further proof of 'state weakness', it is more useful to reflect on the political conditions that produce ineffectiveness. The shortcomings of land registration are the result, in Colombia as mostly everywhere else, of a sort of operational inefficiency, i.e., a set of dysfunctions that create opportunities for rent collection, and are less a sign of administrative weakness than the result of a tacit agreement between a large number of powerful actors on the importance of inaction.

At the same time, all around the world, cadastral registries are part of the usual toolkit used for 'environmental peacebuilding', as they are purported to strengthen societies' capacities to solve land and resource conflicts in a peaceful manner (Pantuliano 2009; Unruh and Williams 2013; Grajales 2020b). In Colombia, the first time a clear nexus was made between the cadastre and peacebuilding efforts was in 2009, when the Constitutional Court called for a reform of the rural cadastral registry, considering that the cadastre's weaknesses were an obstacle to the implementation of a policy protecting the land of the displaced population. This ruling was influenced by the works of the Land Protection Project (the PPTP, see Chapter 4), whose members were already concerned about the consequences of inefficient land registration on IDPs' capacity to hold on to their rights to land.[11] When PPTP members joined the Land Restitution Unit, they strived to push the cadastre problem onto the post-conflict policy agenda.[12] As negotiations on the agrarian chapter opened up in Havana, government officials included cadastral reform in the policy package showcasing the state's commitment towards rural areas. The cadastral registry had become a post-conflict policy.

At the same time, throughout 2016 and 2017, the ANT was designing its intervention programme; its core was a procedure of 'social property planning' (OSP, for *ordenamiento social de la propiedad*). The idea was to put in place a massive territorial intervention during which the ANT would provide an assessment of the territory, analyse property and use rights, and take a series of actions including the formalization of private property, the allocation of public domain lands, and the regularization of titles and plot boundaries. OSP can have a redistributive effect, but it does not necessarily pursue this objective. As an FAO contractor working for the government on the implementation of OSP procedures explained:

> In restitution, it's clear that there's a victim and an alleged perpetrator. In the procedures carried out by the ANT, it is not that clear, because there are two legal subjects and it is necessary to define which one's right should prevail. It is not because the landlord has a lot of land and the peasant has very little that the peasant's claim must be acknowledged. It is necessary to use legal criteria to define who has rights to land and what the nature of those rights entails.[13]

Initially, the cadastre was only one of a variety of technical tools, all pursuing the same objective: solving potential or actual conflicts around land. However, from the onset of the ANT's work, the place of the cadastre in the agrarian agenda progressively grew, until it became the government's flagship policy in 2018. In late 2019, when I interviewed a number of officials in charge of agrarian policies, the cadastre was no longer a tool serving the aims of broader regulation of property rights. It had become a 'technical fix', concentrating the attention of both donors and domestic institutions. But this version of the cadastre had been disconnected from any (re) distributive ambition. It was a mere representation of existing property rights. How did this transformation happen? My take is that a general hostility towards redistribution created a favourable context for a technical approach to land rights, one that insisted on issues of funding or cartography techniques over long-term political objectives. In this case, the technique worked as an 'anti-politics machine' (Ferguson 1994). My research found two main trends in terms of depoliticization.

First, as CRR was translated into sectoral routines, hierarchies between different policy issues were determined by the differences of power and influence between the organizations in charge of them. Beginning in 2016, the (re)distributive aspects of the policy were put in the hands of the ANT, then a new and still poorly staffed agency. The ANT simply did not have the technical capabilities or the available personnel to participate in discussions on the cadastre. Moreover, the ANT was made to wait until May 2017 for a legal instrument (Decree No. 902) to finally establish a framework regulating its operations. At the same time, the technical aspects of the cadastre were put in the hands of the DNP, an organization that is home to some of the finest and better trained economists and policy analysts in the entire Colombian public sector. The DNP is an extremely well-connected agency and enjoys a reputation of professionalism and technical expertise among foreign donors. Fairly logically, the DNP centred the whole discussion around the technical and financial aspects of the intervention, carefully avoiding the most politically sensitive questions.

During the last year of the Santos administration, rivalries within the administration and disappointment from foreign donors led to a deadlock on any discussions regarding the cadastre. From the outset, Duque's administration sent the message to the donors that the cadastre was an important priority for the agrarian agenda.

Technical agencies such as the ANT were sidestepped and a presidential advisor was put in charge of the matter. A number of legislative reforms related to the cadastre were pursued and loan agreements for a USD 150 million funding package were signed with the World Bank and the Inter-American Development Bank between August and October 2019.

However – and this is the second trend of depoliticization – this new dynamic came at the price of a cost-killing attitude intended to generate new momentum and to achieve the greatest possible objectives. This involved detaching the cadastre from most OSP operations – which, because they involved various legal procedures, were seen as more expensive – as well as from community work. In November 2019, the government announced its target for the joint World Bank and IDB project: 72 municipalities were covered by the project, but only 13 of them would be subject to an OSP procedure. When I asked both the World Bank and the DNP project managers about that choice, they gave me the same explanation: the cost for a cadastre operation varied between 140 and 170 US dollars per plot, while an OSP operation required an additional 500 US dollars.[14]

While the cadastre stood at the top of the government's agrarian agenda, OSP operations did not disappear. It is difficult to be categorical about this aspect, as most of these operations were only beginning at the time of my last research stay, but the priorities set by the ANT insist on the necessity of delivering quantitative results. One of the agency's contractors explained to me:

> Initially, we had five municipalities in our zone, but we removed one of them because when we checked the land registries, we realized that there were so many problems… We would have had to initiate seizure and allocation procedures. It was too much! We might have been able to deliver something like 70 titles. That would have been a mess, because it was well below our target. That is why our policy has been to focus on places where problems are simple.[15]

This does not seem to be an isolated case, as other interviewees insisted on the fact that ANT management has instructed their teams to systematically avoid complicated procedures and to focus on zones where a maximum number of plots can be covered with minimal resources. This, of course, is a general trend observed by many analysts working on comparable land registry policies (see, for instance Labzaé 2019). However, in Colombia, monitoring indicators emphasizing the number of plots delivered irrespective of the identity of the beneficiaries and the transformative effects of the titling, result in one particular outcome: they provide a technical – and ostensibly apolitical – justification for excising redistribution from the agrarian policy agenda.

Conclusion

As we have seen, from the outset of CRR, interests opposing redistribution were powerful and well organized. However, this chapter has insisted on the fact that, while the governing elites that rose to dominance after the 2018 general elections are fiercely opposed to any idea of redistribution, the political intentions of a small – although influential – group of people do not explain everything. To understand the betrayal of the Havana promises one must enter into the machine of public policies and finely observe the mechanisms through which land redistribution has been replaced by a technical definition of agrarian policies.

To that end, I traced in the first section how a window of opportunity for transformation opened up between the beginning of the peace talks in 2012 and the 2014 general elections. A momentum for change was used by a broad network of policy actors, located in state agencies and donor institutions, to push a transformative agrarian agenda. As it was designed, CRR was far from revolutionary, but could be considered as a first step towards a new appraisal of agrarian issues.

In spite of this momentum, the window for change rapidly closed. This was the result of an interaction between various trends: the weakening of the Santos government, the polarization around the peace talks brought about by the strategies of the radical right, and the broad mobilization of the political elite in favour of a conservative stance that was protective of the proprietarian social order. The chapter then moved on to analyse the Duque government's output during its first year in power. While the fears of many development professionals, who foresaw a coordinated dismantlement of the entire CRR policy, had not proven themselves founded at the time of writing, I showed that a more discreet but purposeful transformation took place during this period. Agrarian policies were downscaled in the political agenda, seasoned practitioners were circumvented by the new administration, and foreign donors were confined to a less influential policymaking position.

In the final section of the chapter, I went into a more detailed assessment of the transformations in policymaking mechanisms, tracing how the redistributive dimension was excised from two policy tools: the land fund and the cadastre. In each case, the redistributive promises of the Havana agreement were broken when broad objectives were translated into specific administrative procedures and indicators. The key point here is that the demise of redistribution was not done upfront through highly visible political gestures. It was undertaken through the everyday implementation of administrative decisions and designs. The promises of land redistribution were not killed; they were left to die.

The next chapter will return to my Magdalena case study to ascertain what changed between the early 2000s era of high-intensity violence (dealt with in chapters 2 and 3), and the supposed post-conflict transition of the 2010s. As the demobilization of paramilitary militias and the implementation of land restitution promised peace and change, I will contend that the dominant trait revealed upon careful examination is the consolidation of warless capitalism.

Notes

1 Interview with Álvaro Balcázar, Bogota, November 2019.
2 Interview, Bogota, October 2017.

3 Interview with the former coordinator of the report, Bogota, November 2019.
4 Of course, disagreement existed. Humanitarian action coordinating officers working for several Scandinavian donors, who had traditionally been very concerned by human rights violations, tended to be more sceptical about this transition. However, a presentation of these divergent positions falls outside the scope of this chapter.
5 Interview, Bogota, July 2015.
6 Interview, Bogota, October 2017.
7 Interview, Bogota, former ANT head of legal division. November 2019.
8 Interview, Bogota, EU programme manager. November 2019.
9 Interview, Bogota, October 2017 and November 2019.
10 Interview, Bogota, November 2019.
11 Interview, former PPTP and URT official, Bogota, October 2017.
12 *Ibid.*
13 Interview, Bogota, November 2019.
14 Interviews, Bogota, November 2019.
15 Interview, Bogota, November 2019.

References

Acuña, F. and N. Pérez-Guevara. 2019. Desempeño electoral de la FARC en las elecciones de 2018. *In*: F. Barrero, ed. *Elecciones presidenciales y de congreso 2018: nuevos acuerdos ante diferentes retos*. Bogotá: Fundación Konrad Adenauer, pp. 201–41.

Aguilera Peña, M. 2014. *Guerrilla y población civil: trayectoria de las FARC 1949–2013*. Bogotá: Centro Nacional de Memoria Histórica.

Allain, M. 2016. Défendre le territoire: la construction de solidarités internationales par les organisations paysannes colombiennes. PhD diss. Université de Bordeaux, Centre Emile Durkheim, Bordeaux.

Basset, Y. 2018. Claves del rechazo del plebiscito para la paz en Colombia. *Estudios Políticos*, (52), 241–75.

Beltrán, W.M. and S. Creely. 2018. Pentecostals, Gender Ideology and the Peace Plebiscite: PGN, 2020 Colombia 2016. *Religions*, 9(12), 418.

Berdal, M. and D. Zaum, eds. 2017. *Political Economy of Statebuilding: Power After Peace*. London: Routledge.

Bermúdes, A. 2013. Agro Godo Seguro. *La Silla Vacía*, 27 Oct.

Bermúdes, A., ed. 2018. *Los debates de La Habana: una mirada desde adentro*. Bogotá: IFIT.

Bermúdes, A., ed. 2019. *La fase exploratoria del proceso de paz: una mirada desde adentro*. Bogotá: IFIT.

Botero, S. 2017. El plebiscito y los desafíos políticos de consolidar la paz negociada en Colombia. *Revista de ciencia política*, 37(2), 369–88.

Campbell, S., D. Chandler, and M. Sabaratnam. 2011. *A Liberal Peace?: The Problems and Practices of Peacebuilding*. London: Zed Books.

Carranza-Franco, F. 2019. *Demobilisation and Reintegration in Colombia: Building State and Citizenship*. London: Routledge.

Castañeda, D. 2014. *The European Approach to Peacebuilding: Civilian Tools for Peace in Colombia and Beyond*. London: Palgrave Macmillan.

Celis, L. 2018. *Luchas campesinas en Colombia (1970-2016): resistencias y sueños*. Bogotá: Desde Abajo.

Chernick, M.W. 2008. *Acuerdo posible: solución negociada al conflicto armado colombiano*. Bogotá: Ediciones Aurora.

Elias, N. 2001. *The Society of Individuals*. New York, NY: Continuum.
Ferguson, J. 1994. *The Anti-Politics Machine: Development, Depoliticization, and Bureaucratic Power in Lesotho*. Minneapolis, MN: University of Minnesota Press.
García Trujillo, A. 2020. *Peace and Rural Development in Colombia. The Window for Distributive Change in Negotiated Transitions*. London: Routledge.
Garibay, D. 2005. De la lutte armée à la lutte électorale, itinéraires divergents d'une trajectoire insolite. Une comparaison à partir des cas centraméricains et colombien. *Revue internationale de politique comparée*, 12(3), 283.
Grajales, J. 2019. Les terres de la paix. Politiques de l'aide, politiques foncières et sortie de conflit en Colombie. *Gouvernement et action publique*, 8(4), 25–47.
Grajales, J. 2020a. A Land Full of Opportunities? Agrarian Frontiers, Policy Narratives and the Political Economy of Peace in Colombia. *Third World Quarterly*, 47(7), 1141–60.
Grajales, J. 2020b. Losing Land in Times of Peace: Post-War Agrarian Capitalism in Colombia and Côte d'Ivoire. *The Journal of Peasant Studies*.
Grajales, J. and C. Jouhanneau. 2019. L'ordinaire de la sortie de guerre. *Gouvernement et action publique*, 8(4), 7–23.
Gusfield, J. 1989. Constructing the Ownership of Social Problems: Fun and Profit in the Welfare State. *Social Problems*, 36(5), 431–41.
Gutiérrez, F. 2019. The Politics of Peace: Competing Agendas in the Colombian Agrarian Agreement and Implementation. *Peacebuilding*, 7(3), 314–28.
Gutiérrez, F. and M. Marín. 2018. Tierras en el posconflicto: ¿en el fondo cuál es el problema? *Análisis político*, 31(92), 18–38.
Labzaé, M. 2019. 'La Terre est au Gouvernement': droits fonciers, encadrement bureaucratique et conflictualité politique dans deux périphéries éthiopiennes. PhD diss. Université Panthéon-Sorbonne, Paris.
Leech, G. 2011. *The FARC: The Longest Insurgency*. London: Zed Books.
McGee, R. and I.G. Heredia. 2012. Paris in Bogotá: The Aid Effectiveness Agenda and Aid Relations in Colombia. *Development Policy Review*, 30(2), 115–31.
Ottaway, M. 2002. Rebuilding State Institutions in Collapsed States. *Development and Change*, 33(5), 1001–23.
Palier, B. 2005. Ambiguous Agreement, Cumulative Change: French Social Policy in the 1990s. *In*: W. Streeck and K. Thelen, eds. *Beyond Continuity: Institutional Change in Advanced Political Economies*. Oxford: Oxford University Press, pp. 127–44.
Pantuliano, S., ed. 2009. *Uncharted Territory: Land, Conflict and Humanitarian Action*. Rugby: Practical Action Publishing.
PGN (Procuraduría general de la Nación). 2020. *Segundo informe al Congreso. Sobre el estado de avance de la implementación del Acuerdo de Paz 2016–2020*. Bogotá.
Pugh, M.C., N. Cooper, and M. Turner, eds. 2008. *Whose Peace?: Critical Perspectives on the Political Economy of Peacebuilding*. Basingstoke: Palgrave Macmillan.
Ramírez, M.C. 2011. *Between the Guerrillas and the State: The Cocalero Movement, Citizenship, and Identity in the Colombian Amazon*. Durham, NC: Duke University Press.
Rein, M. and D. Schön. 1996. Frame-Critical Policy Analysis and Frame-Reflective Policy Practice. *Knowledge and Policy*, 9(1), 85–104.
Restrepo, J.C. and A. Bernal. 2014. *La cuestión agraria: tierra y posconflicto en Colombia*. Bogotá: Penguin Random House.
Rettberg, A. 2020. Peace-Making Amidst an Unfinished Social Contract: The Case of Colombia. *Journal of Intervention and Statebuilding*, 14(1), 84–100.
Reyes, A. 2013. *La reforma rural para la paz*. Bogotá: Debate.
Richani, N. 2002. *Systems of Violence: The Political Economy of War and Peace in Colombia*. Albany, NY: SUNY Press.

Richmond, O. 2012. *A Post-Liberal Peace*. London: Routledge.
Richmond, O.P. and R. Mac Ginty. 2015. Where Now for the Critique of the Liberal Peace? *Cooperation and Conflict*, 50(2), 171–89.
Roa-Clavijo, F. 2021. *The Politics of Food Provisioning in Colombia: Agrarian Movements and Negotiations with the State*. London: Routledge.
Rodríguez Lach, A. 2019. Una paz austera. *Dejusticia*, 12 Oct.
Rodríguez-Ruíz, M.J. 2019. ¿Se difuminó la bancada de la paz?: un análisis sobre el comportamiento legislativo de los partidos políticos en el marco del 'Fast Track' para la implementación del acuerdo de paz. B.A. Dissertation. Universidad Javeriana de Colombia, Bogotá.
Schön, D.A. and M. Rein. 1995. *Frame Reflection: Toward the Resolution of Intractrable Policy Controversies*. New York: Basic Books.
Suhrke, A. and M. Berdal. 2013. *The Peace In Between: Post-War Violence and Peacebuilding*. London: Routledge.
Surel, Y. 2000. The Role of Cognitive and Normative Frames in Policy-Making. *Journal of European Public Policy*, 7(4), 495–512.
UNDP, (United Nations Development Programme). 2011. *Colombia Rural: Razones para la Esperanza. Informe Nacional de desarrollo Humano 2011*. Bogotá.
Unruh, J. and R. Williams. 2013. *Land and Post-Conflict Peacebuilding*. London: Routledge.
Vallejo, M. and C. Montoya. 2017. *Los cercos del debate sobre restitución de tierras: Encuadres retóricos de la Ley 1448 de 2011 en la prensa colombiana nacional y regional*. Bogotá: Editorial Pontificia Universidad Javeriana.
Verdad Abierta, V. 2016. Los cuestionamientos a bananeros detrás del No. *Verdad Abierta*, 13 Oct.
Zamosc, L. 1986. *The Agrarian Question and the Peasant Movement in Colombia: Struggles of the National Peasant Association, 1967-1981*. Cambridge: Cambridge University Press.

6

TOWARDS WARLESS CAPITALISM

Violence levels in Magdalena – as in most paramilitary strongholds – diminished in the latter half of the 2000s, primarily as a result of the partial demobilization of militia groups between 2003 and 2006. In 2010, a total of 307 murders were recorded in the department (ODH 2014). This was the lowest tally in two decades. The drop in violence was even more staggering when looked at from the point of view of internal displacement. While 47,205 people were registered as IDPs in Magdalena in 2002, there were only 2421 internal refugees in 2010 (ODH 2014). By 2012, the land restitution policy (see Chapter 4) had begun to be implemented. A land restitution unit (URT), tasked with documenting cases of violent dispossession and bringing them before a judge – thus acting as the victims' legal counsel – was created. Particularly in the period immediately following its creation, the unit was staffed with people who had spent years working with victims of forced displacement and who were therefore seen by this vulnerable public in a positive light (Dávila 2018). In Magdalena, the first regional director of the URT had a long history of working for human rights NGOs and had previously worked on several cases of land dispossession on the northern Caribbean coast. I had met him a few years before, in 2009, when he was coordinating a programme for victims of forced displacement and violent dispossession at a prominent Bogota-based NGO.

By 2019, during the final research trip I went on for this book, a land formalization programme related to Comprehensive Rural Reform (CRR) was being implemented in northern Magdalena. The project was headquartered in the town of Ciénaga, and its operations were primarily focused on the mountain areas of the lower Sierra Nevada, where many small peasants lack property titles. These are precisely the areas where, three decades before, the first paramilitary violence entrepreneurs had prospered thanks to the drug economy. The rural reform in Magdalena, as in many areas of paramilitary domination, has been reduced to these formalization programmes. These are considered as pacified territories where peacebuilding policies have a limited reach. This final chapter will study some of the characteristics shared by these regions, which are now at the forefront of post-conflict development schemes. Through a careful analysis of the transformations affecting economic accumulation, I will outline the characteristics of this 'post-conflict' inequality regime: warless capitalism.

In most societies, land access is determined by a combination of market forces, political power, and direct coercion (Hall *et al.* 2011). In times when armed violence is a common means of social interaction, direct coercion can be the dominant method, without necessarily superseding the other two. As a matter of fact, situations of protracted violence like those I have described in previous chapters are characterized by the use of violence both as a way of gaining control over property, and of consolidating power over the institutions that recognize property rights (Lund 2016). Conversely, warless situations are characterized by a variation in the uses of violent methods. While armed violence cannot be expected to disappear altogether, its use typically becomes less visible and widespread. This is not solely – nor even primarily – due to the effectiveness of peacemaking policies, but rather, more commonly, because dominant actors are able to transform their violent domination into a more institutionalized configuration, both through political institutions and through the market (Aspinall 2009; Woods 2011; Ybarra 2012; Martiniello 2013; Lund 2018).

How do transformations in the use of violence affect access to land? This is the question asked in this chapter. I will return to my local case – Magdalena – studied in the second and third chapters and analyse how economic domination has been fashioned by transformations in the use of violence. I will show that one critical feature of this warless capitalism is the transition from extra-economic accumulation – i.e., based on coercion – to economic accumulation – relying primarily on the mechanisms of the market. This does not mean that violence completely vanishes. On the contrary, selective violence remains an important method for consolidating control over land; most importantly, the *potential* that violence might be used is a pervasive element hampering people's capacity to mobilize and claim their rights, as recent memories of paramilitary brutality cast a shadow over any attempts to collectively organize.

I will argue here that the transition to relatively limited practices of violence has played a fundamental role in the transformations that the capitalist system has undergone. As a regime of accumulation, capitalism has historically been based on a clear-cut distinction between politics and the economy. While the former is a field of deliberation, the latter is supposedly driven by market forces (Polanyi 1944; Rosanvallon 1999; Mitchell 2011). One of the foundations of this differentiation is the fact that accumulation purportedly depends on 'non-authoritative, non-political means' (Meiksins Wood 1995, 29): i.e., a well-functioning market. Of course, the representation of the market as an autonomous social entity is a political illusion. Particularly – but certainly not exclusively – in cases like mine, accumulation is not 'non-authoritative', as coercion is never far away. In reality, market mechanisms often conceal the central role of brute force in the development of capitalism (Klein 2008). But this illusion is extremely powerful, and institutions and discourses that conceal the role of violence in processes of accumulation play a critical role in the development of capitalism (Harvey 2005). What I call warless capitalism is precisely a regime of accumulation that depends on positions of power that were once achieved through violent means that have now been subdued by post-conflict institutions, discourses, and economic forces (Grajales 2020).

This chapter endeavours to describe the consolidation of warless capitalism, using Magdalena to illustrate the dynamics present in other areas formerly dominated by paramilitary forces. In the first section of this chapter, I will demonstrate that the reproduction of a proprietarian social order is no longer primarily achieved through violent means, but now passes instead through a combination of institutional and market forces. In the second section, I will show that social groups that formerly represented a threat to dominant classes are no longer capable of mobilizing in the name of land redistribution.

Warless dispossession

The most visible aspect of Magdalena's social landscape, echoing similar situations all over Colombia, is a highly exclusive inequality regime that requires far less violence to be reproduced than before. Impoverished peasants are excluded from access to land, not primarily through violence, but as a result of the combined effects of institutional and market forces. This has the important consequence of concealing the political dimension of exclusion: people are no longer victims who might be entitled to a remedy, they are subjects unsuited to the requirements of the market. The two following subsections will provide two examples of this.

Surrounded by the plantation

In Chapter 3, I described the case of the Tierra Grata community,[1] which had been forced to leave its land due to a combination of paramilitary coercion and economic vulnerability. The last members of the community were displaced in 2004 after being compelled to leave by paramilitary commander Carlos Tijeras. In parallel with this violent land grabbing, the Serranos – a family of rich agro-industrialists from the area – seized control of the land. The fact that the peasant community was deeply in debt allowed the property transfer to be completed lawfully. The bank foreclosed on the estate and auctioned it off, an opportunity presumably seized upon by the Serranos. Since then, the new owners have claimed that they were unaware of the circumstances under which the people of Tierra Grata were forcibly displaced.

The case of Tierra Grata exemplifies the transformations in the forms of land grabbing and dispossession that took place in the decade following the paramilitary aggression perpetrated against the community. The start of the decade was characterized by a decline in direct violence, as the partial demobilization of paramilitary groups reduced the grip that their networks held over social life. However, more discreet forms of violence remained, as former members of paramilitary militias sold their services as violence entrepreneurs to powerful local figures. During the early 2010s, enhanced public scrutiny over land dispossession resulted in increased efforts to conceal the violent means used to acquire or consolidate private property. This was done through institutional leverage and market relations.

In July 2005, Carlos Tijeras, the paramilitary leader responsible for evicting the Tierra Grata community from its lands, was arrested in the city of Barranquilla. The militia formerly placed under his command, the William Rivas Front, laid down its weapons in March 2006 as a part of the AUC's demobilization process. Motivated by these events, some members of the displaced community chose to return to the land that they still considered as their own. After a few months of collective organization, 12 families entered the estate on 1 December 2006 with the intention of taking back their lands. The new owners regarded the peasants as squatters and immediately requested police intervention. Between December 2006 and April 2007, several unfruitful police-led eviction attempts were made. The community was able to resist law enforcement. On 18 April, armed men, presumably working as security contractors, seized control of the estate. Three bulldozers were used to lay waste to the camp where the peasants lived and the few hectares of oil palm trees that they had managed to plant. These men, who claimed to work for a business partner of the Serranos, the Daabon group, were actually former paramilitary militia members. It was never fully clear whether they had set up a legal security firm – as some of them claimed – or if they were simply one of the numerous gangs that flourished in the aftermath of the AUC's demobilization.

The characteristics of the two business conglomerates that claimed to have exclusive legal rights over the estate underscore the daunting challenge faced by the peasant community. The Daabon group was then one of Colombia's biggest palm oil producers. At the time this event occurred, it directly exploited 4000 hectares of palm oil plantations and had business agreements with other producers representing an additional 2000 hectares. In addition, Daabon held significant investments in banana and coffee plantations and owned several service providers linked to the agribusiness sector. Daabon had greatly benefitted from the Colombian government's policy in favour of agro-fuels and had become a pioneer in the transformation of palm oil into agro-diesel.[2] By 2009, Daabon was running the largest agro-diesel plant in Colombia. In parallel, it had established itself as the Colombian leader on the organic palm oil market and had become the number-one provider for the British retailer, The Body Shop. In 2010, The Body Shop cancelled its contract with Daabon in response to the Colombian conglomerate's ties to a land-grabbing scandal that was revealed in a report commissioned by *The Guardian* and London-based NGO Christian Aid (Syal and Brodzinsky 2009, 2010). The Serranos themselves were in charge of a powerful agribusiness conglomerate, with more than 2000 hectares of northern Magdalena farmland among their holdings. In the 1990s, they became influential figures in the region's booming palm-oil sector. One of the Serrano brothers also served in the Colombian House of Representatives between 2012 and 2018.

The political and economic influence of these two conglomerates would not facilitate the task for the peasants of Tierra Grata. At the same time, the political exposure of both the Daabon group and the Serrano family became a constraint for their actions, particularly in the late 2000s and early 2010s, when the issue of violent land grabbing began to receive more prominent media coverage and attention from politicians. In 2006, the Serrano family and Daabon signed a partnership for the exploitation of palm oil plantations in Tierra Grata. Over the next three years, about 300 hectares of the total surface of 500 hectares would be covered by the plantation. Both the peasants and the corporations resorted to legal means in their efforts to emerge victorious. The peasant community sought protection from the authorities and claimed that their land had been violently grabbed and subsequently appropriated by Daabon and the Serranos. The two corporate partners filed several complaints against the peasants for trespassing and squatting. During this period, the peasant community split into two rival factions: one willing to negotiate with the new landlords and another that continued to engage in its struggle for the land. The former not only abandoned all new disputes with the corporate alliance running the estate, but it became instrumental in the legal conflict that unfolded after 2007. Within the context of this protracted legal battle, they offered support for the landlord's complaints and testified that the occupants had harassed the company's workers in an attempt to prevent them from working at the plantation.

In 2009, for reasons that I was unable to discern in my research, Daabon withdrew from the business partnership and the Serrano family alone assumed the exploitation of the estate. By that time, the entire estate had been utterly transformed by the palm-growing activities. A group of some 20 families settled in the areas of the estate that were not planted with palm. While most of the violent harassment stopped (although the legal quarrels continued), the everyday life and livelihoods of the community were hindered by the way in which the estate had been reconfigured. In the words of one of the community's members, who is still struggling today:

> We went to the area where there were no palm trees, because we knew that we could not touch the trees if we wanted to stay out of trouble. But we were surrounded by palm trees; the managers of the estate closed off the roads so we couldn't pass. Then we received support from an NGO, the Justice and Peace Commission, who helped us open the road. They helped us make our situation public knowledge. This was important because one of our opponents was a congressman, so he was sensitive to his public image.[3]

Map 6.1 shows the reconfiguration of the landscape. As mentioned by my interviewee in the above quote, the main way of reaching the estate was to pass through the plantation zone. In addition, some of the neighbouring estates are controlled by large landowners who are seldom sympathetic to the peasant community's struggle. One of them, the Chimborazo estate (to the northeast of Tierra Grata), was even the site of egregious physical and sexual violence in the late 1990s, and even today,

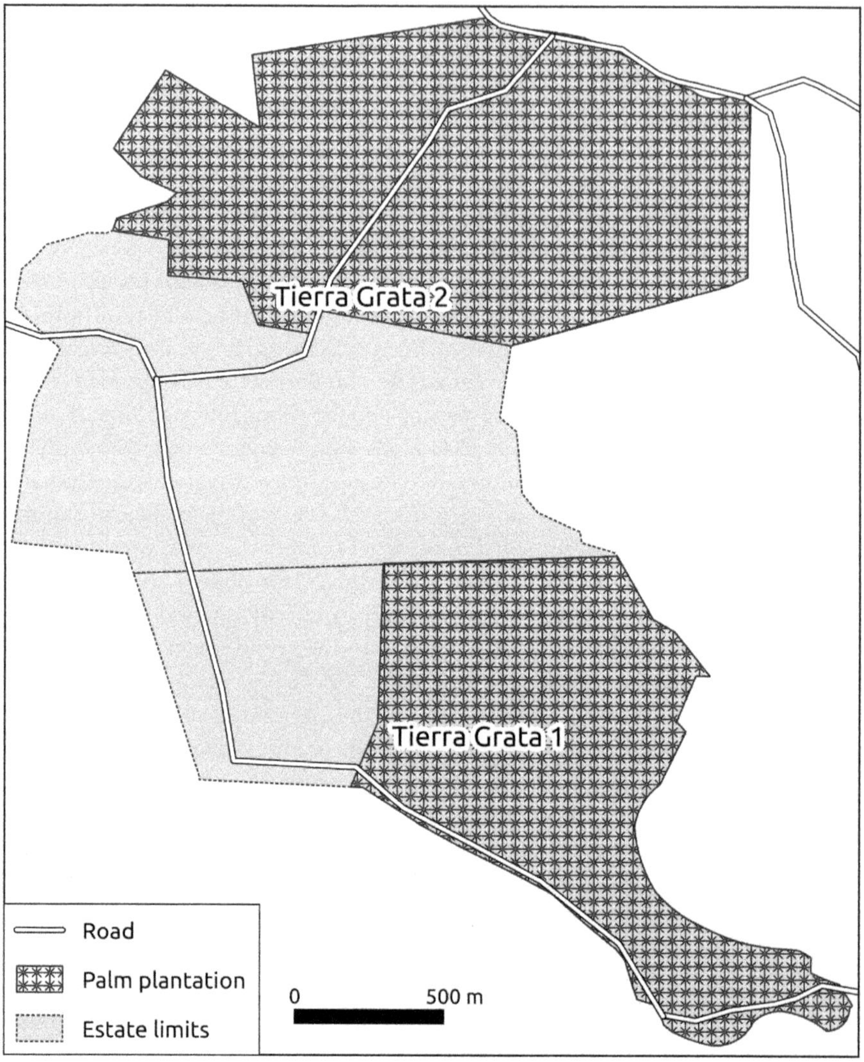

MAP 6.1 Surrounded by the palm plantation in Tierra Grata. Source: the author, created using data from the Colombian cartographic and cadastral authority IGAC (Licence CC BY 4.0)

the peasant community that was forced to leave it claims to be the true owner of the estate.

In addition to the transformation of the landscape, the remaining members of the peasant community have been consistently undermined in their efforts by a combination of institutional and economic forces. While they are now able to farm the edges of the palm plantation, they are not recognized as the legal owners of their land. Consequently, they are unable to access credit or public subsidies, and they have met with enormous difficulties in their efforts to pursue an economically viable farming venture. In the words of one of them:

> Today, we essentially grow food crops for our own sustenance. We have struggled to begin farming some cash crops as well. First, we managed to plant a papaya crop at the cost of a tremendous effort, but the trees became diseased and we didn't have the money to pay for the treatments. Now I have begun to prepare a seedbed for coffee and bananas, but I don't have the money to start production.[4]

Economic hardship has led several peasants to sell some or all of their land to people outside the original community. Of the 20 families who were still active on the estate in July 2018, nine are newcomers. These are landless peasants who benefitted from the complicated situation of the Tierra Grata community to purchase land at fire-sale prices. Their arrival has made it more difficult to find a collective solution to the conflict, as most of them hope to reach individual settlements.

The legal struggle itself has been undermined by different factors linked to the vulnerable situation of the community. Ten families dropped their lawsuits in 2014. They belonged to the original beneficiaries of the collective land title, but had fled due to paramilitary aggression and conflicts with the new landlords. The remaining members of the community today believe that these others were probably offered money. This is not the only legal obstacle to the peasants' claims. In July 2014, the leadership of the peasant group signed a memorandum of understanding with one of the companies in the Serrano conglomerate. The document stated that neither the association nor its members had been submitted to pressure, harassment, or threats from the Serranos or their affiliates. This was likely a legal strategy used by the landowner to impede any further legal action – particularly criminal charges. In exchange for the approval of the memorandum, the peasants received 15,000 US dollars to be shared among them. When I asked today's president of the association why they had signed this document, he told me:

> First, because we were starving. With that money we were able to support ourselves for a few months and we used it to plant the papaya seedlings. And second, because we don't have any evidence against this man. Nothing that ties him to the paramilitaries. So we know that the land he has already planted with palm is lost. Now, all our efforts are aimed at holding on the portion that he hasn't planted yet.[5]

The case of Tierra Grata shows how the forms of exclusion and access were transformed in the years following the demobilization of paramilitary groups and the increase in the visibility of the problem of land grabbing. When the community returned to the estate in 2006, violence was on the decline due to the recent demobilization of the paramilitary militias. However, armed threats remained, as many former militiamen had moved into the business of violence entrepreneurship and continued to sell their services to well-paying patrons. Over the next few years, the use of overt violence became more costly, particularly due to the involvement of prominent members of the patrician establishment in the estate's agribusiness project. This does not mean that the peasants' lives became easier. Various legal, institutional, and economic obstacles hindered their claims. Their dreadful living conditions made collective action more difficult, and some members of the community became weary of their struggle and abandoned their claims altogether, motivated by the meagre pay-offs offered by their opponents or by buyers willing to purchase their uncertain rights over the land. Overall, the history of Tierra Grata is emblematic of the impact of the transformations that have taken place in the realm of security. As analysed in a different Colombian setting by Teo Ballvé (2020), while post-conflict transition hardly equates to peace, it introduces a new set of constraints and opportunities for dominant economic actors. Both peasants and corporations have learned to play this new game. But just as in times of war, they are not competing on equal terms.

Landscapes of dispossession

This example illustrates the fact that, in warless times, forms of domination that formerly required high levels of violence to be sustained tend to be naturalized and inscribed in the social order. One aspect of this process is geography. Political geographer Diana Ojeda has made an important contribution by demonstrating that we 'need to understand the processes and dynamics of dispossession from its spatialities, that is, from the violent forms in which they are inscribed and sedimented in space, and which result in their production' (Ojeda 2016, 20). Thinking in terms of 'landscapes of dispossession' sheds light on the fact that land grabbing is not a circumscribed event but a continuous social process resulting in a 'socio-environmental reconfiguration' (Ojeda *et al.* 2015). This theoretical model is particularly important for my study.

Between 2018 and 2020, in collaboration with William Renán and Benjamin Lévy and with the support of an environmental engineer, I coordinated a cartographic survey, using remote sensing, and covering two of the banana district's river basins, those of the Río Frío and the Sevilla rivers (Grajales *et al.* 2020). Our curiosity had been piqued by our field research in the area, as peasants had reported that the main factor that was driving them out of their lands was no longer violence, but a lack of water. During the dry months from January to March and July to August, rainfed agriculture had become impossible, and further sources of water were desperately needed. However, especially in the lower-lying areas of the region, streams were dry at this period of the year. Our interviewees told us stories of lost crops, hunger, and crushing debt, but also of digestive problems and skin conditions caused by them being forced to use polluted water for drinking and bathing. And yet, irrigation canals are present throughout the area – many of which had been built decades before in the years of the United Fruit Company. When the corporation left the area, these canals were taken over by the Colombian Institute

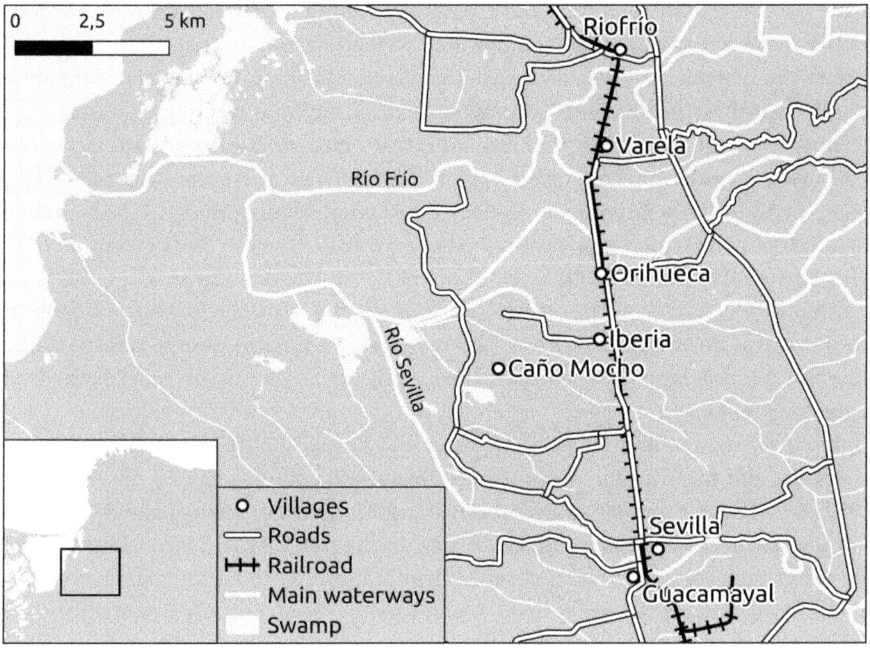

MAP 6.2 Area of the cartographic survey

Note: only the main rivers, streams, and swamps are represented. *Source*: the author, created using data from IGAC (Licence CC BY 4.0)

for Agrarian Reform (Incora), which was supposed to use them to encourage the development of small- and mid-sized farming. However, the fact that many of these canals are dry during half of the year is the result of the fact that water is captured upstream by more powerful actors with better political connections.

Our aim was to understand the causes and processes behind water grabbing. The two areas covered by the study correspond to a total surface area of 14,000 hectares (Map 6.2). This zone comprises a dual property structure: commercial farms of over 20 hectares (18 per cent of the properties) are present alongside small exploitations of less than 5 hectares (47 per cent of the plots) (DANE 2016). However, the distribution of property in the region is not homogeneous. Land is much more concentrated in eastern areas of the banana district, which are located near the highway leading to Santa Marta and the railroad, and it tends to be more fragmented further west. This east–west divide also corresponds to the course of the region's rivers, which come down from the Sierra Nevada mountains in the east and run towards the swaps in the west. As a consequence, when the weather is dry, the larger commercial farms located upstream are able to grab most of the water before it reaches the lower zones.

With no government regulations to rely on, vulnerable actors are powerless to resist water grabbing. Notwithstanding this property structure, water grabbing is a matter of force, both economic and coercive. All along the banks of the rivers, landowners have set up pumping facilities used to feed their own irrigation infrastructure at the expense of other users. An interviewed business lawyer who frequently works with these firms opined that most of these facilities lack the necessary legal permits to pump water, but that the authorities have seldom taken any action to stop them. In addition, landlords frequently hire armed personnel to protect their water facilities. Several peasant leaders I interviewed in July 2018 had chilling encounters with armed men while trekking on half-dry river beds taking pictures to document the situation. In early March 2020, as the dry season was at its peak, a group of 15 peasants went into an estate to ask the manager to stop pumping, as no water was flowing downstream. The manager – a masked, assault-rifle-wielding man in typical paramilitary attire standing at his side – summarily ordered them to leave the premises.

By analysing satellite imagery, my team was able to locate the canals and water reservoirs and measure the evolution of this infrastructure over two decades, from 1997 to 2018. Our findings were entirely consistent with the results of field research and the testimony of peasants from the area. In the past 20 years, the total surface of the water reservoirs in the area climbed from 27 km^2 to 183.4 km^2, and the length of the irrigation canals grew from 13.21 km to 77.68 km. Our estimations are that the total capacity of the irrigation system in 2018 was over 1.5 million m^3 of water, compared to only 207,660 m^3 in 1997. This means that the amount of water extracted from the rivers has risen sevenfold in just the last 20 years.

Climate change and weather phenomena such as El Niño, which have resulted in longer, dryer summers, and irregular rainfall, have only aggravated the situation. Furthermore, river flows have been affected in recent decades by deforestation in

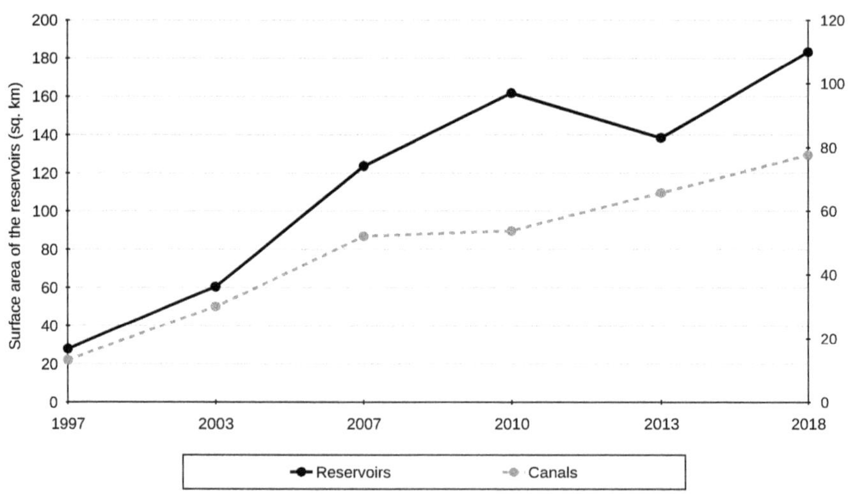

FIGURE 6.1 Growth of irrigation systems in the Río Frío and Río Sevilla basins.
Source: the author

the Sierra mountains and the expansion of coffee production. Lacking sufficient measurements, official environmental agencies are unable to produce estimates of the problem, but several environmental scientists I met with at the University of Magdalena consider that the river flows from the Sierra mountains have significantly diminished.

As multiple phenomena have rendered water a scarcer resource, the consequences of the privatization of this vital liquid have become even more severe. They have an impact not only on the livelihoods of local inhabitants, but also on fragile ecosystems. The Ciénaga Grande wetlands are one of these natural areas. Considered as a sanctuary for endemic flora and fauna and protected by the Ramsar Convention on Wetlands, their ecological balance depends on the flows of freshwater from the rivers and saltwater from the sea. As the former decrease, these marshes have fallen prey to a process of salinization, with dreadful consequences for plant and animal life.

These changes in the environment have further reinforced the consequences of water grabbing. When the rivers and the canals downstream dry up, the only remaining source of this vital liquid is groundwater. However, its overexploitation, combined with the salinization of the marshes, has come to result in the salinization of groundwater. For several months of the year, this brackish water is the only thing available for many people. When used for drinking or farming – and depending on its levels of salinity – this water can result in negative consequences for human, animal, and plant health.

When seen from a broader perspective, the privatization of water is a further step in the transformation of agrarian capitalism. Small-scale agriculture, which relies on public or shared access to water, has been further hindered by these transformations. As a result, many small peasants are selling their land and leaving, adding themselves to the mass of rural migrants already in the cities. In the village of Caño Mocho, for instance, those I interviewed estimated in July 2018 that a third of the population had already left. Those who stayed were struggling to survive as food and cash crops perished due to the lack of water. However, most of the peasants I interviewed did not perceive the situation as equal to dispossession. Land is paid for at market prices, which are fairly good, as the development of agribusiness in the area attracts new flows of investments. Once investors gain access to new land, they generally have the necessary resources and political connections to obtain water as well. While silent and discreet, the result of these transformations is the progressive disappearance of peasant agriculture and a further consolidation of a corporate model of land ownership. Here, the market conceals the power relations that support the existing inequality regime.

Water is only one example of how the privatization of natural resources spurs social and agrarian change. The same village of Caño Mocho illustrates how water grabbing interacts with the confiscation of other commons. In interviews with the villagers, one of the recurring topics was the transformation of the surrounding landscape, as banana plantations have now come to dominate the area encircling the village. According to my interviewees, one fundamental change in their way of life accompanied the expansion of neighbouring plantations in the early 2010s. Previously, access to a neighbouring swamp provided them with firewood, fish, and bushmeat. This swamp came to be fenced off by the corporation that owns most of the neighbouring plantations. According to cadastral registers I was able to consult, this was public land, and its use by a private corporation should theoretically be considered unlawful. However, not only did the company grab and fence off the land, it also drained the swamp and tore down the trees and bushes to expand its plantation. In practical terms, the loss of access to this resource meant that the villagers now needed to purchase food and fuel that they had previously extracted from nature. As a consequence, they have become completely dependent on the salaries earned by young men and women on the plantations or in the city.

The cases of Tierra Grata and Caño Mocho are related to the violent history of Magdalena in different ways. While this relationship is fairly straightforward in the former case, it is much more difficult to ascertain in the latter. This reflects the difficulties encountered by many researchers when they try to link physical violence and structural violence. Of course, one can consider that both forms of violence lie along a continuum of methods used to maintain social order. This is a common argument in scholarly literature, made both by Galtung's (1969) classic works and by more recent scholarship (e.g., Farmer 2001; Scheper-Hughes and Bourgois 2003). However, when confronted with real-life situations, social scientists often have a hard time proving causal relations and singling out the mechanisms tying these various forms of violence together. My view is that two possibilities exist that allow us to respond to this methodological challenge. The first is to trace these relations in multiple cases and pick out a pattern. This is what I did in the third chapter of this book. This could be done with water grabbing as well. Many companies now massively purchasing land and investing in irrigation infrastructure have a share of their investments in lands that were violently grabbed in the early 2000s. For instance, the company that owned Las Franciscas (see Chapter 3) at the date of my last visit is a subsidiary of one of the largest agribusiness groups in the region, which has played a pivotal role in the process of water privatization. However, a detailed study of these relationships exceeds the scope of my data. I do not have access to a clear understanding of the business relationship between corporations, investors, and holdings. As commercial registers conceal the real owners behind complicated legal schemes designed to avoid scrutiny, much more research would be necessary to understand the ties between the current-day corporate structure of the banana business and the history of violence in the region.

Consequently, here I will use a second strategy – doubtless less ambitious but more empirically robust – to show the link between structural and physical violence. My argument in the following section will be that the trajectory of brutal armed violence, as well as land restitution's incapacity to transform the fundamental structures of economic domination, has hindered most possibilities of social protest and collective action that existed in previous decades. This demonstrates a direct connection between past physical violence and present structural violence.

Restitution and 'legal security'

In 2018, as the Santos government reached the close of its term, the administration was asked to look back at the achievements of its restitution policy. Interviewed by a news magazine, the general director of the URT claimed that 80 per cent of the restitution requests had been processed, and that the institution was well on its way to achieving the objectives set by the law in 2011 (Semana 2018). A closer look at URT's work shows a much more nuanced and less optimistic image. The unit's reach is limited in both its scope and its purpose. This, much research has shown, is not due to administrative weakness – a rather simplistic explanation often used for policy shortcomings in the Global South. On the contrary, the URT was home to an efficient, transparent, and very committed body of public servants (Dávila 2018). The policy's drawbacks should instead be seen as an inherent part of the very way it was designed. Land restitution was never conceived as a policy aiming to radically transform the countryside, but rather as a moderate intervention, justified in the name of transitional justice, allowing Colombian society to 'turn the page' from dispossession (Carranza *et al.* 2019; Gutiérrez 2019). Given the way in which it was intentionally diluted, one can effectively consider that the land restitution policy as enacted fulfilled its promises.

The first section showed the multiple ways in which accumulation and dispossession have been reconfigured by the decrease in violence, the partial demobilization of paramilitary groups, and the greater visibility of the issue of violent land grabbing. One aspect of this problem will retain our attention in the remainder of the chapter. In 2011, some hoped that the Victims Act would result in deep transformations in regions such as Magdalena, where gunpoint capitalism had become the norm over the previous two decades. By December 2020, the policy had resulted in the restitution of 384,753 hectares of land, concerning 8931 plots nationwide.[6] While the individual effects of the law are undeniable, the land restitution policy has not transformed the landed inequality regime. I will sketch an analysis of this limited structural impact in the following pages by focusing on two important points: first, the near absence of criminal liability provided for in the restitution policy, which has protected most land grabbers from criminal proceedings; and second, the very modest reach of the restitution policy and the ways in which it has contributed to naturalizing a representation of regions such as Magdalena as peaceful territories.

Restitution without liability

A good indicator of the built-in moderation of the transitional justice system in Colombia is how it purposefully sidestepped the problem of business actors' criminal liability. This is due not only to a strategy implemented by a variety of powerful actors in Congress and the executive branch, but also to the everyday organization and functioning of investigative bodies. This shortcoming broadly affects the institutional framework designed for paramilitary demobilization (Justice and Peace

Act, 2005), the 2011 Victims Act, and the measures adopted under the 2016 Havana peace agreement.

The Justice and Peace Act designed a two-track procedure aimed respectively at paramilitary commanding officers and rank-and-file troops. The former were processed through a special criminal jurisdiction run by the attorney general's office and endowed with significant resources and legal authority. The latter were subject to a non-judiciary transitional justice procedure intended to gather their testimony as part of a truth and reconciliation programme. However, both procedures focused on the individual liability of ex-combatants, and did not consider the analysis of larger networks to be a priority. This is particularly true regarding the ties between paramilitary groups and business actors, even if, very early in the process, critical voices and scholarly research called for more severe action against paramilitary business associates (Romero 2011).

According to Wesche (2019), who has undertaken the most thorough inquiry on the subject to date, third-party criminal liability is limited by the very design of the procedure. Whenever a prosecutor gathers information on the involvement of business actors in paramilitary networks, she is legally obligated to transfer the case to a different division, as the special unit in charge of paramilitary officers has no jurisdiction over third parties. The investigation is then transferred to a local attorney general's division, which inevitably has more limited resources and a heavier caseload. These local prosecutors have no incentive to pursue such cases, which are generally seen as complicated, difficult to document, and even dangerous for their lives. As local prosecutors are evaluated on quantitative indicators, investigations that have poor chances of delivering immediate results are often dismissed. This, according to Wesche, explains the very low impact of criminal investigations against business actors. By 2018, the author identified 376 criminal proceedings open against business actors, 85 per cent of which were still in a pre-trial phase (Wesche 2019, 488).

Proceedings against rank-and-file militiamen are even less concerned with the criminal liability of business actors. To the best of my knowledge, there has not been any thorough scholarly research into these procedures. Consequently, I can only rely on my own – rather exploratory – research on the subject. The contribution to justice, truth, and reconciliation by militiamen was assigned to the National Commission for Historical Memory (CNMH). Its Truth Agreements Division was in charge of interviewing demobilized combatants and gathering the information in databases used to produce the CNMH's various publications. The legal framework involved greatly limited use of the testimony, as the law forbade using the information delivered through these mechanisms to further criminal investigations. In addition, the regional CNMH divisions, at least in the case of Magdalena, were particularly concerned by the legal risks involved in collecting and compiling compromising information. In July 2018, I met with former CNMH agents who had been in charge of gathering ex-combatants' testimony. One of them told me that:

> The lawyers at the [CNMH] regional division put many obstacles in front of us. They were very careful with any information concerning third parties, especially when it involved influential businessmen and corporations. They told us not to enter this information into the databases. According to them, even if the databases could not be used in court, there was the risk that someone who was mentioned in the testimony would sue the CNMH for overstepping its jurisdiction.[7]

The 2011 Victims Act added a further layer to Colombia's transitional justice infrastructure. However, it did nothing to address the problem of business actors' criminal liability. As with the Justice and Peace Act, the institutions in charge of restitution policy are simply required to forward any information on potential criminal behaviour to the attorney general's office. Surprisingly, while the law's point of departure was the observation of the massive land grabbing and violent dispossession that took place within the context of armed conflict, no provisions were included that would have made it possible to disentangle the responsibilities at play behind the economic motives of war.

There are two primary reasons for this. The first was the necessity for the government to promote a non-antagonistic definition of the restitution policy, thus excluding the much more controversial issue of criminal liability from the debate (Vallejo and Montoya 2017; Gutiérrez 2019). As a matter of fact, powerful members of the agrarian elite organized a campaign in opposition to the law, arguing that it would 'dispossess' honest and hardworking farmers based on unproved allegations. Magdalena was one of the core places where these mobilizations took place, spearheaded by the ranchers' association. The second reason is procedural. One of the main innovations of the Victims Act is an expectation of 'good faith with due diligence'. This a legal doctrine under which current owners of disputed land need to prove not only that they did not directly participate in land grabbing, but also that they acted with 'extra care or diligence' in making sure that the property they were purchasing was not the product of dispossession. This means that local URT divisions, which are in charge of collecting information on land restitution claims, have to meet standards of proof that are far below those of criminal courts. In other words, evidence gathered to support a restitution claim is often insufficient to pursue a criminal investigation. This logic is fully endorsed by members of the institution, who consider that their work is to deliver restorative justice, not to reveal who is legally responsible for dispossession. One of the lawyers interviewed at the Santa Marta URT division told me:

> Land restitution is a civil law procedure, not a criminal investigation. My job is not to establish that Mr X or Y was an ally of paramilitary groups and that they helped him drive a peasant community from their lands. My job is to prove that this peasant community has a legal right to restitution and that the current owner of the land failed to act with the necessary diligence to be

> entitled to compensation. That's all. If I have information about the criminal networks behind this, I forward it along to the attorney general's office.[8]

The most recent addition to the Colombian transitional justice system is derived from the Havana agreements. The peace accord created a Special Jurisdiction for Peace (JEP), which became extremely controversial as the Santos coalition broke apart in 2016, and has now become a favourite target for right-wing politicians. The agreement initially provided for a much more comprehensive system than the Justice and Peace Act had. The JEP was created not just to uncover the crimes committed by the FARC, but, more broadly, as a special court with jurisdiction over any crime committed within the context of the armed conflict. As such, unlike the institutions already in place, the JEP brought with it the potential to prosecute civilians who had supported and benefitted from paramilitary activities (Freeman and Orozco 2020). However, the agreement's translation into effective legislation stripped the JEP of its jurisdiction over business actors' liability (Wesche 2019, 498–500). During the negotiations and the accord's ratification process, this particular point was violently attacked by the right-wing opposition parties, chief among which were former President Uribe's Centro Democrático and the Conservative Party. It also came under fire from members of the Santos' coalition, including former vice-president Germán Vargas Lleras and his party. As the ruling coalition was weakening at the approach of the 2018 general election, people like Vargas Lleras opportunistically saw attacking the most controversial aspects of the peace agreement as a promising electoral platform. In the end, the JEP ultimately ended up being toothless in cases involving the economic sponsors and business partners of armed groups. While the law does not exempt businesspeople from the JEP's jurisdiction, they can only be judged if they voluntarily submit themselves to it.

The quasi-absence of any enforcement of business actors' criminal liability helps explain why many former paramilitary associates remain prosperous businesspeople today. Criminal charges have only been pursued when the actions of business associates have gone well beyond the economic sphere, as in the case of Augusto Castro, a rich Magdalena rancher who had come to play an instrumental role in the AUC's military and financial apparatus (Quinche Roa *et al.* 2018). Still, these cases required a great deal of pressure from civil society, well illustrated by the conviction of 16 palm oil entrepreneurs who had violently grabbed land belonging to Black communities in the Pacific lowlands (Vargas Reina 2016). In most cases, those who have done business with paramilitary groups have been left alone and allowed to thrive.

The lack of criminal prosecution is an important factor in explaining the persistence of the land inequality regime. Large landowners and agribusiness entrepreneurs who benefitted from the militias' help in expanding their wealth might face the threat of losing some of their estates through restitution requests, but they face only a very slight risk of being made to answer for their crimes in criminal court. They might lose some of their assets, but there is little risk that their structural domination in the political and economic arenas will ever be substantially affected. Moreover, in addition to this quasi-immunity from criminal prosecution, an analysis of the land restitution policy's results to date reveals that it has probably benefitted only a very small share of those potentially eligible for it. The next subsection briefly sketches some hypotheses.

Restitution's limited reach

When viewed from a micro-level, the gap between the realities of land grabbing and the reach of Colombia's restitution policy is staggering. The municipality of Zona Bananera saw 27,088 people forcibly displaced between 2000 and 2013 (ODH 2014). Estimates are that two-thirds of them were peasants (Renán *et al.* 2007). However, only 381 individual restitution requests had been filed as of July 2018, and a mere 85 cases had actually made it to court. Only one class action case (covering 48 people) had been granted a judicial restitution order as of the time of my writing (URT 2018). This is consistent with the findings available in the literature. To the extent of my knowledge, the only quantitative study on those who did not apply to the restitution policy was conducted in 2017 by researchers from Colombia's National University. Professor Francisco Gutiérrez and his team conducted a statistical analysis using the 'multisystem estimation' technique in five municipalities on the Caribbean coast. For these five locations, the probabilities of a land dispossession case actually being reported to the authorities ranged between 4 and 16 per cent (Gutiérrez *et al.* 2018).

Several hypotheses can be drawn from the literature to explain the disparity between the realities of dispossession and its administrative translations. Lack of knowledge is likely responsible for many cases in which the restitution policy is not taken advantage of. A 2014 survey of a representative panel of 498 victims of land dispossession established that 26 per cent of them were unaware of the existence of the restitution policy (Gutiérrez *et al.* 2014). Those who knew of the programme were not necessarily keen on embarking in a long and exhausting process to recover a piece of land they had not seen in decades. The same survey even found that 70 per cent of the panel did not want to return to the countryside. To interpret this result, one must bear in mind that years – or even decades – have passed between the moment at which these forced displacements occurred and the implementation of the law. Among the cases studied in Chapter 3, many potential beneficiaries are now over 60 years old. For them, returning to their lands might mean settling down in places where basic public services, transportation, and health facilities remain difficult to access. Even when the direct beneficiaries did respond that they had a desire to return to the countryside, younger family members who had grown up in the city were seldom attracted by this perspective. Moreover, IDPs have grown accustomed to state inertia and have learned that they can seldom rely on the support of government agencies. This conundrum is well summarized by the daughter of a late land claimant:

> The concern we have is that all the restitution claimants in our case are seniors. Two of them have died in just the last year. Twenty days ago, a man

> from my community died and now my dad [...] has passed away as well. As we see the old people dying, the youth are getting discouraged. The one thing that discourages us the most is that the state has not made the slightest effort to help us get back to our land. It's as though they wanted us to give up.[9]

In addition to the lack of knowledge and the lack of motivation to file a restitution request, there is an even more fundamental factor that affects people's capacity to make a restitution claim. Most of the cases I have studied – in particular those from the banana district – share certain common traits. The majority of the claimants never left Magdalena, and some even stayed in close proximity to the dispossessed land. Community ties have endured, meaning that when the restitution procedure was initiated, organized collective action was possible. This contrasts with what we know about forced displacement and land dispossession. In Magdalena, as elsewhere, forced displacement patterns were characterized by long-distance migration and rupture in community and even family links (Renán 2007). While much more detailed research would be necessary to fully understand the mechanisms in place, it is possible to conclude that land restitution primarily benefitted a minority of people who were able to make themselves 'visible' to government actors.

This visibility is highly dependent on the URT's policies. In the early years of the restitution programme, significant effort was expended to reach vulnerable populations, document violent land grabbing, and understand the patterns of dispossession. In Magdalena, the very first members of the URT's staff had thorough prior knowledge of local dispossession cases. The programme's waning political capital and human and financial resources have changed this, as bureaucratic routines and quantitative indicators have become dominant (Gutiérrez 2019). This transformation at a local level is in reality tied to broader national forces, and it is related to the URT's necessity to showcase rapid results. When a purported victim presents a land restitution request to the unit, the first administrative step is to assess the case for eligibility; URT staff then begin an investigation to document the case and decide whether it can be taken to a judge. According to a research by the Colombian Commission of Jurists – a highly respected human rights NGO – during the early years of its existence (2012–2014), the URT's criteria proved rather lenient, with an admission rate of 60 per cent. Between 2015 and 2017, however, the procedure became much more selective, as the admission rate plummeted to just 29.5 per cent (CCJ 2018, 31). Rejections at the admission phase are based solely on the information provided by the victim. If the narrative is not deemed credible or it does not fit the legal criteria, a URT agent dismisses the demand. The victim's only recourse is to file a complaint in court. The characteristics of this early phase disadvantage vulnerable and less legally fluent victims, who are probably the most invisible as well. As a matter of fact, the URT reported in 2017 that there were only 68 pending complaints, compared to the over 21,055 dismissal orders filed by that time (CCJ 2018, 48). The unit presented this as evidence that dismissal orders were well grounded, but this could also be seen as a sign that dismissed claimants were simply too vulnerable to pursue any further action.

On most of the Caribbean lowlands, where paramilitary rule is now supposedly a thing of the past, contemporary policy action is fashioned by the idea that the countryside has been pacified and the restitution policy has fulfilled its promises. A July 2018 decree terminated the period for filing restitution claims in a majority of the former paramilitary strongholds, including most of Magdalena. This executive decision had already been anticipated by local URT managers. According to URT agents and former agents that I met with in Santa Marta in July 2018, the unit's local policy was being driven by conviction that the core of their jobs had already been accomplished and that they could now move on to a much more bureaucratic and legalistic approach. By the time I was conducting my field research, the staff numbers for personnel tasked with documenting cases, keeping in direct contact with victims, and responding to their needs had been radically reduced, and they had now been supplanted by lawyers whose sole aim was to follow up on pending cases. One of my interviewees – a URT agent – bitterly regretted this transformation: 'I am planning to leave soon, as this job no longer corresponds to what I was looking for. Now, we are basically tasked with keeping a good performance rate and getting judicial decisions out.'[10] The victims themselves realize that the unit has changed: 'Most of the people that I trusted [at the URT] lost their contracts. I no longer know anyone,' one land claimant told me. 'We peasants feel that we no longer have any allies in the state,' she pursued.[11]

A similar attitude was widespread among NGOs, most of whom had played a critical role in previous years by working to spur the state to action and obtain quicker results. As donors' attention has shifted to the former FARC strongholds of southern Colombia, resources have been reallocated. The NGOs' departure has been bitterly felt among land claimants, who fear that the state will no longer listen to their demands. One of them told me:

> NGOs are leaving. The Norwegian Refugee Council is about to leave because they say that their mission here is done. The problem is that without them, no one listen to us. When you go to a public office with a guy wearing a vest with the logo of one of those organizations, they immediately welcome you. If you show up alone, no one cares.[12]

The state's and most NGOs' behaviour reflects the belief that Magdalena has become a pacified place where their action is no longer a priority need. Even if NGO practitioners might feel that sources of violence and exclusion remain, they still believe that their limited resources can be better utilized elsewhere. 'Much remains to be done here, but ours is an organization specialized in humanitarian emergencies. And there is no longer a humanitarian emergency here', an NGO employee told me in July 2018. 'Plus, there are places where our presence is much more urgent',[13] he continued. At the time of our meeting, he was spending his last few days in Santa Marta. He had been transferred to the Catatumbo, a rural region on the Venezuelan border that was on its way to becoming one of the most violent areas in the country.

This poses intractable questions for development and humanitarian aid professionals. While security conditions in Magdalena have indeed improved, the depiction of the region as a pacified place naturalizes the consequences of violence and conceals the mechanisms of exclusion. In addition, this representation critically reduces the resources that victims of violent land grabbing have at their disposal in their struggles to achieve recognition and restitution. After being ignored for more than a decade before going on to become the object of fierce political debate in the late Uribe years, to only afterwards be thrust back into the spotlight by the Santos administration, victims of dispossession have now simply been pushed into oblivion.

Conclusion

In this chapter, I have shown that a real decrease in the scale of violence has not brought about the transformation of the inequality regime that was consolidated during the war. On the contrary, the transformations of the 'post-conflict' era tend to conceal its violent origins and to further naturalize and consolidate an exclusionary social order. My focus has been on the transition between a situation in which direct coercion is a critical method of accessing land, to a more institutionalized configuration in which political and economic institutions – and most prominently the market – become the principal vectors of access and exclusion. This, I have claimed, is not only compelling for the analysis of my specific case, but more broadly for the study of the political economy of peacebuilding. As a matter of fact, as post-conflict institutions and discourses tend to conceal the role of violence in past processes of accumulation, they contribute to one of the most fundamental philosophical foundations of capitalism: the illusion that accumulation in market societies is primarily achieved through non-coercive methods.

My demonstration was divided into two phases. The first section traced the transition from coercive accumulation to a context characterized by more restricted practices of violence. I showed how institutional and market forces combined to sustain an inequality regime that requires much less violence than before to be reproduced. Legal and economic institutions are closely intertwined in the production of these much less visible forms of exclusion: in my case, legal vulnerability weakened peasants' capacity to create economically viable business ventures within a context in which peasant agriculture is marginalized, with smallholders finding themselves surrounded by large-scale commercial plantations. Exclusion is further naturalized by the reconfiguration of the landscape, as water and other commons are grabbed by dominant actors. While the privatization of these commons is carried out rather discreetly, these transformations are characteristic of the progressive erosion of peasant agriculture and the consolidation of a corporate agribusiness model. However, this causation is not direct, as peasants are not coercively pushed off of their lands. They are driven out of agriculture by drought, and many of them consider themselves lucky to be able to sell their lands to agribusiness investors. Again, the market hides the fact that many of these investors are precisely the people responsible for the ecological transformation that pushes peasant farmers to sell.

The second section tied together these various forms of structural violence to past tactics that relied on physical violence. This was done by showing how the legal framework of land restitution, which promised to do away with the legacies of gun-point capitalism, failed to address the most violent traits of the existing inequality regime. Two aspects illustrate this. The first is the absence of any specific concern for holding land accumulators criminally liable for their actions. As this shortcoming concerns the entire Colombian transitional justice system, it explains the nearly complete impunity from which former paramilitary business associates have benefitted. While the AUC militias laid down their weapons and many of their leaders spent several years in prison, very few of their economic sponsors and business partners have been prosecuted. The second aspect is an internal contradiction inherent in the restitution processes, as their admittedly meagre results have nevertheless contributed to fostering an impression that former paramilitary strongholds have become pacified territories where peacebuilding action is no longer needed. While individual actors are fully cognizant of the simplistic assumptions underlying this image, the policies of both the state and civil society organizations have come to be determined by this influential portrayal.

This, I have contended, has resulted in extremely complex dilemmas. True, Colombia today is a more pacified society than in the late 1990s and early 2000s, when the expansion of guerrilla warfare and the growth of paramilitary militias nurtured a state of ceaseless bloodshed. This is particularly true in places like Magdalena, which have had less exposure to the most recent wave of armed violence that has struck former FARC strongholds. These representations of stability and emergency drive the attention of peacebuilding and post-conflict development professionals away from places like Magdalena. To a certain extent, this is a perfectly normal process, the natural outcome of the fact that scarce resources need to be directed where they are most needed. However, by the same token, peacebuilders ironically end up contributing to the concealment of the very violence that they once struggled to reveal. I will address this dilemma in the conclusion of the book.

Notes

1 Sources for this case study are listed in Chapter 3.
2 On this policy, see Grajales (2013), Marin-Burgos (2014), Marin-Burgos and Clancy (2017).
3 Interview, village of Guacamayal, July 2018.
4 *Ibid.*
5 *Ibid.*
6 Forjando Futuros, a Bogota-based NGO, publishes a real-time monitoring of the policy's implementation. See www.forjandofuturos.org.
7 Interview, Santa Marta, July 2018.
8 Interview, Santa Marta, July 2018.
9 Interview, Fundación, July 2018.

10 Interview, Santa Marta, July 2018.
11 Interview, Fundación, July 2018.
12 *Ibid.*
13 Interview, Santa Marta, July 2018.

References

Aspinall, E. 2009. Combatants to Contractors: The Political Economy of Peace in Aceh. *Indonesia*, (87), 1–34.

Ballvé, T. 2020. *The Frontier Effect: State Formation and Violence in Colombia*. Ithaca, NY: Cornell University Press.

Carranza, F., A. Gómez, C. López, and W.E. Ospina. 2019. Límites en el diseño de la restitución en Colombia. *In*: F. Gutiérrez, R. del P. Peña-Huertas, and M.M. Parada-Hernández, eds. *La tierra prometida: Balance de la política de restitución de tierras en Colombia*. Bogotá: Editorial Universidad del Rosario, pp. 39–73.

CCJ, (Comisión Colombiana de Juristas). 2018. *Cumplir metas, negar derechos*. Bogotá: CCJ.

DANE, (Departamento Administrativo Nacional de Estadística). 2016. *Tercer censo nacional agropecuario*. Bogotá: DANE.

Dávila, J. 2018. A Land of Lawyers, Experts and "Men without Land": The Politics of Land Restitution and the Techno-Legal Production of "Dispossessed People" in Colombia. PhD diss. Harvard University, Harvard.

Farmer, P. 2001. *Infections and Inequalities*. Oakland, CA: University of California Press.

Freeman, M. and I. Orozco. 2020. *Negotiating Transitional Justice: Firsthand Lessons from Colombia and Beyond*. Cambridge: Cambridge University Press.

Galtung, J. 1969. Violence, Peace, and Peace Research. *Journal of Peace Research*, 6(3), 167–91.

Grajales, J. 2013. State Involvement, Land Grabbing and Counter-Insurgency in Colombia. *Development and Change*, 44(2), 211–32.

Grajales, J. 2020. Losing Land in Times of Peace: Post-War Agrarian Capitalism in Colombia and Côte d'Ivoire. *The Journal of Peasant Studies*. DOI: 10.1080/03066150.2019.1691535

Grajales, J., W. Renán, and B. Lévy. 2020. *La gobernanza del agua en la zona bananera del Magdalena. Concentración del agua como mecanismo de exclusión económica y social*. Lille; Santa Marta: University of Lille; University of Magdalena, Working paper.

Gutiérrez, F. 2019. Lo bueno, lo malo y lo feo de la restitución de tierras en Colombia: una lectura política e institucional. *In*: F. Gutiérrez, R. del P. Peña-Huertas, and M.M. Parada-Hernández, eds. *La tierra prometida: Balance de la política de restitución de tierras en Colombia*. Bogotá: Editorial Universidad del Rosario, pp. 9–37.

Gutiérrez, F., P. García, and C. Argoty. 2014. *La restitución y sus problemas según sus potenciales beneficiarios*. Bogotá: Observatorio de Restitución y Regulación de Derechos de Propiedad Agraria.

Gutiérrez, F., M. Marín, P. Perdomo, and D. Machuca. 2018. Arañando la superficie: subestimaciones sistemáticas en la política de restitución y sus fuentes. *Estudios Socio-Jurídicos*, 20(1), 133–77.

Hall, D., P. Hirsch, and T.M. Li. 2011. *Powers of Exclusion: Land Dilemmas in Southeast Asia*. Honolulu, HI: University of Hawaii Press.

Harvey, D. 2005. *The New Imperialism*. Oxford: Oxford University Press.

Klein, N. 2008. *The Shock Doctrine: The Rise of Disaster Capitalism*. London: Penguin Press.

Lund, C. 2016. Rule and Rupture: State Formation through the Production of Property and Citizenship. *Development and Change*, 47(6), 1199–228.

Lund, C. 2018. Predatory Peace. Dispossession at Aceh's Oil Palm Frontier. *The Journal of Peasant Studies*, 45(2), 431–52.

Marin-Burgos, V. 2014. Access, Power and Justice in Commodity Frontiers: The Political Ecology of Access to Land and Palm Oil Expansion in Colombia. PhD diss. University of Twente, Enschede.

Marin-Burgos, V. and J.S. Clancy. 2017. Understanding the Expansion of Energy Crops Beyond the Global Biofuel Boom: Evidence from Oil Palm Expansion in Colombia. *Energy, Sustainability and Society*, 7(21).

Martiniello, G. 2013. *Accumulation by Dispossession, Agrarian Change and Resistance in Northern Uganda*. Kampala: Makerere Institute of Social Research (MISR), Makerere University.

Meiksins Wood, E. 1995. *Democracy Against Capitalism: Renewing Historical Materialism*. Cambridge: Cambridge University Press.

Mitchell, T. 2011. *Carbon Democracy: Political Power in the Age of Oil*. London: Verso.

ODH (Observatorio de derechos humanos de la vicepresidencia de la República de Colombia). 2014. *Estadísticas de violaciones a derechos humanos y Derecho Internacional Humanitario (1990-2013)*. Bogotá: Vicepresidencia de la República.

Ojeda, D. 2016. Los paisajes del despojo: propuestas para un análisis desde las reconfiguraciones socioespaciales. *Revista Colombiana de Antropología*, 52(2), 19–43.

Ojeda, D., J. Petzl, C. Quiroga, A.C. Rodríguez, and J.G. Rojas. 2015. Paisajes del despojo cotidiano: acaparamiento de tierra y agua en Montes de María, Colombia. *Revista de Estudios Sociales*, (35), 107–19.

Polanyi, K. 1944. *The Great Transformation: The Political and Economic Origins of Our Time*. New York, NY: Farrar & Rinehart.

Quinche Roa, J.M., P. Perdomo Vaca, and J. Vargas Reina. 2018. *El despojo paramilitar en el Magdalena: el papel de las élites económicas y políticas*. Bogotá: Observatorio de Restitución y Regulación de Derechos de Propiedad Agraria.

Renán, W. 2007. *Contextualización del desplazamiento en el departamento del Magdalena y Santa Marta (1997-2007)*. Santa Marta: Universidad del Magdalena, Vicerrectoría de Investigación, Fonciencias, Sección I, informe final convocatoria 2004.

Renán, W., J. Barbosa, and W. Suárez. 2007. *La propiedad rural en el Magdalena 1970-2004 y algunas relaciones con el desplazamiento forzado*. Santa Marta: Universidad del Magdalena, Vicerrectoría de Investigación, Fonciencias, Sección II, informe final convocatoria 2004.

Romero, M., ed. 2011. *La economía de los paramilitares*. Bogotá: Random House Mondatori.

Rosanvallon, P. 1999. *Le capitalisme utopique. Critique de l'idéologie économique*. Paris: Seuil.

Scheper-Hughes, N. and P. Bourgois. 2003. *Violence in War and Peace: An Anthology*. Malden, MA: Wiley-Blackwell.

Semana. 2018. "Nuestro legado será haber cambiado el fusil por un juez": Ricardo Sabogal. *Semana*, 27 Feb.

Syal, R. and S. Brodzinsky. 2009. Body Shop ethics under fire after Colombian peasant evictions. *The Guardian*, 12 Sep.

Syal, R. and S. Brodzinsky. 2010. Body Shop drops supplier after report of peasant evictions in Colombia. *The Guardian*, 2 Oct.

URT. 2018. Unidad de restitucion de tierras. Territorial Magdalena. Subcomité de tierras. Municipio Zona Bananera. Unpublished document.

Vallejo, M. and C. Montoya. 2017. *Los cercos del debate sobre restitución de tierras: Encuadres retóricos de la Ley 1448 de 2011 en la prensa colombiana nacional y regional*. Bogotá: Editorial Pontificia Universidad Javeriana.

Vargas Reina, J. 2016. Despojo de tierras paramilitar en Riosucio, Chocó. *In*: F. Gutiérrez and J. Vargas Reina, eds. *El despojo paramilitar y su variación: quiénes, cómo, por qué*. Bogotá: Universidad del Rosario, pp. 121–46.

Wesche, P. 2019. Business Actors, Paramilitaries and Transitional Criminal Justice in Colombia. *International Journal of Transitional Justice*, 13(3), 478–503.
Woods, K. 2011. Ceasefire Capitalism: Military–Private Partnerships, Resource Concessions and Military–State Building in the Burma–China Borderlands. *The Journal of Peasant Studies*, 38(4), 747–70.
Ybarra, M. 2012. Taming the Jungle, Saving the Maya Forest: The Military's Role in Guatemalan Conservation. *The Journal of Peasant Studies*, 39(2), 479–502.

CONCLUSION

Over the last decade, the term 'dispossession', *despojo*, has become a buzzword in Colombian politics. As it was introduced by activists, jurists, and scholars to refer to the massive, violent land grabbing of peasant lands by paramilitary groups and their allies, it eventually became a matter of legal debate, a category whose limits, characteristics, and content had to be determined by the law. But 'dispossession' was also the subject of much political struggle. It was used by those who demanded a more adamant commitment from the state in favour of forcibly displaced peasants who were faced with tremendous challenges when returning to their lands. It was also abusively used by the most regressive landed elites, who claimed that they themselves were now on the brink of becoming victims of a new form of dispossession, as the state was threatening to grab back the lands they had accumulated through violence or deception.

This book argues that the term 'dispossession', while politically and analytically meaningful, does not fully cover the complexity of the land issue, neither in Colombia, nor in many other comparable cases. In particular, it fails to grasp the complex ties between armed violence and land conflicts. Unequal access to land – far from being the mere result of an exclusionary political project – is embedded in the formation of politics and the market. In this respect, three elements play a fundamental role over the long term: the political marginalization of the peasantry, the production of agrarian frontiers, and the strong territorial fragmentation of the state, which most often operates through the privatization of control over resources and populations. Beyond the apparent disruptions of cycles of violent confrontation and appeasement, these three elements remain constants in Colombian contemporary history.

Against the backdrop of this diagnosis, I have pointed to a fundamental contradiction in the economic peacebuilding policy package. Indeed, I have questioned the premise – which has widespread traction even well beyond Colombia's borders – of the possibility of differentiating between outright dispossession and legitimate accumulation in a post-war conjuncture. The belief in a neat distinction between legitimate and ill-acquired assets is the result of two factors. First, it is hard-wired into the mechanisms of post-war policymaking; in fact, it is related to yet another illusion – namely, the existence of a clear distinction between times of war and times of peace. This, I have argued, ultimately conceals the continuity that exists between warring and warless economic, social, and political orders. Second, and most importantly, the distinction between legitimate and plundered assets is a necessary belief for many actors in the post-conflict milieu, as it circumscribes a field of action, drawing the limit between the economic problems that need to be addressed (restorative justice, criminal accumulation, etc.) and those issues that are beyond the scope of peacebuilding. By the same token, the differentiation between the problems that can and should be addressed and all the others is common in development circles. It amounts to what one might call a 'doing as if' attitude, which consists of divorcing 'structural sources of inequality from their technical domain' (Li 2007, 275), thus considering that some good can still be done without touching the big, complicated – rather messy – picture.

This 'doing as if' attitude is often required for policymaking to function as it should, for the everyday interactions between government officials, foreign donors, and NGO practitioners, and for the mental well-being of anyone committed to doing a meaningful job in these places. However, despite all the good will and the tireless energy that some members of these policy communities showcase, this book argues that the necessary illusions that render this everyday work possible can ultimately play into the hands of a system that reproduces structural inequality and domination. For many people, turning a blind eye on the egregiously violent foundations of post-war capitalism is not a matter of ignorance, bad faith, or naïveté. On the contrary, it is a necessary trade-off, a hard decision that enables them to make use of the meagre possibilities offered by post-war times and, at the end of the day, to do some good. I myself, as a citizen, a committed observer, and a teacher of young people who will join the ranks of the development and humanitarian aid world, most certainly understand this trade-off. But the duty of scholarly analysis is not to judge people's intentions, but to show that the social world is about much more than these intentions, even when they are profoundly noble (Li 2014). What I have tried to demonstrate is that this 'doing as if' attitude is more than an operational requirement. Under some conditions, it can become a critical part of the process through which power relations previously produced and reproduced in violent ways are laundered and made respectable.

Ultimately, this book intends to place social inequality and economic domination at the heart of the study of the continuum between times of war and warlessness. From this point of view, it engages with a growing and vibrant field of scholarly works that, over the last decade, have critically assessed the economic foundations and implications of peacebuilding and post-conflict development policies. This field draws its origins from a critical assessment of the belief in the capacity of the free market to deliver stability and well-being in the aftermath of war.

The foundational studies in this field focused on demonstrating the authoritarian foundations of the 'liberal peace' project, as well as showing how these assumptions nurtured new forms of accumulation, domination, and social stratification (Pugh *et al.* 2008; Richmond and Mac Ginty 2015; Mac Ginty and Williams 2016; Berdal and Zaum 2017). After this initial phase, the scholarly discussion has moved on to a more in-depth understanding of how everyday economic processes work and how ruling positions and accumulation interact and intersect, concealed under the cover of the comfortable cloak of 'peace' (Cheng 2018; Distler *et al.* 2018; Jennings 2018).

I believe that the Colombian case can provide a meaningful contribution to this discussion because it shows that liberalism, even when it is well intentioned and progressive, is not enough. My study has shown that, in Colombia, liberal peacebuilders have failed to deliver a peaceful and more equal society, not because they are not committed enough, but because liberal thinking is fundamentally curtailed by its belief in the clear-cut distinction between politics and the economy (Rosanvallon 1999). As the former is the realm of deliberation, the latter is governed by market forces. Most often, market regulation is justified by the market itself, in order to eliminate 'distortions'; at most, it can pursue the apparently progressive aim of shaping competition, so that economic actors face each other on equal terms. One of the most fundamental consequences of this differentiation is that it conceals the political foundations of the market (Meiksins Wood 1995), i.e., the fact that the market itself is a product of political forces that reflect a balance of power, a fundamentally unequal relation (Polanyi 1944; Meiksins Wood 2002; Bourdieu 2017). In accordance with this line of thinking, this book argues for a broader understanding of how markets are produced and shaped through violent means, and how the legacies of violence endure well beyond these direct effects. It also argues in favour of a rigorous analysis of how markets work to shroud the power relations that reproduce an inequality regime, offering each of us the comforting conviction that there is nothing wrong with how resources are distributed.

My exploration of these issues began in the first chapter, with a panoramic view of the history of the 'agrarian problem' in Colombia. My aim was not solely to provide the necessary context to readers unfamiliar with the case, but to adopt a narrative covering a larger time frame than that of the contemporary armed conflict. My intent in this chapter was to show that violent accumulation and land conflicts predate and extend beyond the history of the Colombian civil war. I did not intend to find the 'roots' of violent conflict, but rather, to depict the formation of an inequality regime. This inequality regime was, at some points in history and in different manners depending on the region, affected by the use of violence as a means of accumulation. However, it cannot be reduced to these most visible forms of accumulation – plundering and armed dispossession – as violence is only one way, among others, used to sustain and reproduce a brutal social order.

My second chapter traced the ways in which one specific category of armed actors – paramilitary militias – emerged in this pre-existing setting. It tells the story of the transformation of Magdalena's violence entrepreneurs into a new breed of anti-communist militias. It shows how these armed groups became critical security providers for the owning classes of the region less through their capacity to provide protection from the rebel threat than their willingness to become actors of social and political policing. The chapter then moved on to show how the local militias were co-opted by a newcomer – the AUC – and how this group set up close ties with local power holders. One of the consequences of this relationship was the convergence of economic, political, and military interests around violent land grabbing.

The third chapter narrowed the book's analysis to the contentious banana district of Magdalena. I traced the ties in this territory between violence and agrarian capitalism over a period of 15 years. I see this chapter as adding a critical contribution to the book's argument as a whole, as it shows, in detail, the ways in which paramilitary violence was embedded in relations of economic domination concerning land and labour. The chapter also shows how precarious and uncertain the possibilities of land access were over those years for peasants living in areas such as Magdalena. The stories of four struggles for land were described. In these cases, not even the most organized, tenacious, and resourceful communities managed to resist the brutal wave of violence that was unleashed by the alliance between local landlords and paramilitary groups. This chapter tells a story of frustrated hopes and failed mobilizations, connecting these individual stories to the overarching power of class domination.

The fourth chapter brought in the discussion on peacebuilding by tracing the origins of the land restitution policy that arguably marked a clean break with prior treatment the relationship between land and war in Colombia. I trace the origins of this legislation in the history of humanitarian assistance for internally displaced persons, as well as the introduction in Colombia of the tools of transitional justice and the debates surrounding them, on the occasion of the paramilitary group's partial demobilization after 2003. When the 'Victims Act' was passed by Congress in 2011, it was advertised by the administration and by many commentators as a revolutionary step in recognizing the intricate ties between land and war. However, by the same token, the law reduced this relationship to a single dimension: dispossession. I inquire into the reasons for this framing and its underlying mechanisms, showing how the government endeavoured to limit the scope of the law and prevent the debate from extending beyond the circumscribed field of restorative justice. As a result, the law traced a fixed line between criminal spoliation and the forms of inequality and economic domination that formed the very substrate of dispossession.

The fifth chapter opened with the idea that, to some of its supporters, the peace negotiations with the FARC brought with them the promise of opening up the debate on land accumulation and inequality and moving beyond restorative justice towards a redistributive approach. These promises were embodied by the agrarian chapter of the peace agreement, which created a policy of 'Comprehensive Rural Reform' (CRR). Thanks to the commitment of a small group of policy professionals and the support of foreign donors and Colombian NGOs, the CRR policy began under good auspices. However, its promises were quickly watered down as the failure of the ratification referendum and the weakening of the second

Santos administration robbed the idea of real reform of any political steam it may have previously had. The final blow to the promises of the Havana talks was struck with the 2018 presidential elections and the arrival of a socially regressive government to power. The new administration has not dismantled the post-conflict land policy – as many feared it would – but it has made it extremely difficult to attain any meaningful result by slashing its resources, downscaling its political importance, and disarticulating the policy network that created and sustained it.

The sixth and final chapter returned to the local scale, with an account of the transformations of accumulation in today's Magdalena. A former paramilitary stronghold, this territory has seen a relative improvement in its security conditions during the last decade. However, this transformation has not shaken up the inequality regime that was consolidated during the war. Rather the opposite, it has contributed to concealing the violent foundations of the social and economic order. This results from various factors. First, the transition from violent accumulation to milder forms of coercion played a fundamental role in transforming the perception of land grabbing. Second, the shortcomings of transitional justice have resulted in a situation in which not only is the criminal liability of land grabbers not actually addressed, but the illusion of a return to normal is systematically maintained. Overall, both the transformations in the use of violence and the public policies addressing the consequences of the war have contributed to depicting Magdalena – just like other Colombian territories – as a pacified place.

This book's analysis, while centred upon the last few decades, has covered a period of one century, tying armed violence to a longer history of accumulation, economic domination, and dispossession. However, this book is not only about analysing the events of the past. While I am not naive enough to believe that my book can have a concrete impact on the life of those I interviewed in Magdalena and beyond, I do think that it can participate in a broader debate on the future of agrarian policies. Schematically, this debate opposes those who believe that renewed efforts should be made to open new avenues for change, and those who feel that such an endeavour is ultimately pointless.

In December 2017, I had a taste of that debate, as I was invited to participate in a round table on the Colombian peace process organized by the Paris-based Foundation for the Study of the Humanities (FMSH). After I had given a short talk based on some of the ideas that would later go on to nurture this book, I passed the floor to the next speaker, a popular Colombian novelist and essayist. He began his presentation by responding to mine, saying that he found it exaggerated to grant such a pivotal place to rural issues despite 70 per cent of Colombia's population living in the city. In a certain way, his view echoed that of a number of Colombian urbanite intellectuals, who pictured their country as a growing economy that would be better off fighting for the development of high-tech and services than remaining mired in the agrarian quarrels of the previous century. My fellow panellist was not the only visible advocate of this opinion. In December 2014 and January 2015, James A. Robinson, the highly influential and widely read economist and political scientist, wrote two op-eds in a Colombian newspaper arguing a similar position

(Robinson 2014, 2015). Robinson, who has taught for several years as a visiting professor at Los Andes University in Bogota, made a controversial point, which led to much discussion in Colombian intellectual circles. According to him, the only way to solve the country's centuries-old agrarian problem was to 'ignore it and let it wither away'. For Robinson, instead of focusing their efforts on the Colombian countryside, progressive policymakers should endeavour to create opportunities for people in the cities. As a result, impoverished peasants could look for prosperity elsewhere than in the countryside, until the landed elites became 'as unnecessary and anachronistic as the British aristocracy'. While Robinson recognized that building a more equal and fair rural society could be an honourable aspiration, he believed that the obstacles were so daunting that policymakers would be better served by focusing on less antagonistic development prospects such as education and urban development.

Most replies to Robinson's op-eds were focused on the issue of justice. They generally argued that the Colombian state has accumulated a huge debt towards rural territories and peasants. As the signature of the Havana agreement was then a visible prospect, the time had finally come to pay that debt. Critics of Robinson's argument considered that it essentially boiled down to a denial of justice, to considering that the page of violence and dispossession could simply be turned, all the pain and suffering could be forgotten, and all peasants' claims for redress could be dismissed. From my point of view, this book defends a critique that is both broader and more radical. As it stands, I argue that the problem with the Colombian countryside is not simply a matter of justice, it is a matter of power: of who has it and who will have it in the years to come. As I have shown, over the last century, the formation of agrarian capitalism has been deeply intertwined with the transformations of political power. Every form of power that has been used, from institutional influence to brute force, has been harnessed to protect a highly exclusionary social order and, the resulting inequality regime has furnished the bedrock for the elimination of dissent and the repression of progressive alternatives. Nothing indicates that Colombia is moving away from this political climate. As the government widely advertises agribusiness development opportunities and champions massive investment into the primary sector, the power of those who control land and other natural resources is probably larger than ever. As such, addressing the intricate relationship between violence and agrarian capitalism is not only a historical and memorial duty, it is – in Colombia as just about everywhere else in Latin America and the Global South – a fundamental step towards the establishment of a more democratic social contract.

References

Berdal, M. and D. Zaum, eds. 2017. *Political Economy of Statebuilding: Power after Peace*. London: Routledge.

Bourdieu, P. 2017. *Anthropologie économique - Cours au Collège de France 1992-1993*. Paris: Le Seuil.

Cheng, C. 2018. *Extralegal Groups in Post-Conflict Liberia: How Trade Makes the State.* Oxford: Oxford University Press.

Distler, W., E.B. Stavrevska, and B. Vogel. 2018. Economies of Peace: Economy Formation Processes and Outcomes in Conflict-Affected Societies. *Civil Wars*, 20(2), 139–50.

Jennings, K.M. 2018. Peacekeeping as Enterprise: Transaction, Consumption, and the Political Economy of Peace and Peacekeeping. *Civil Wars*, 20(2), 238–61.

Li, T.M. 2007. *The Will to Improve: Governmentality, Development, and the Practice of Politics.* Durham, NC: Duke University Press.

Li, T.M. 2014. Anthropological Engagements with Development. *Anthropologie & développement*, 37–38–39, 227–40.

Mac Ginty, R. and A. Williams. 2016. *Conflict and Development*. 2nd ed. London: Routledge.

Meiksins Wood, E. 1995. *Democracy Against Capitalism: Renewing Historical Materialism.* Cambridge: Cambridge University Press.

Meiksins Wood, E. 2002. *The Origin of Capitalism: A Longer View*. London: Verso.

Polanyi, K. 1944. *The Great Transformation: The Political and Economic Origins of Our Time.* New York, NY: Farrar & Rinehart.

Pugh, M.C., N. Cooper, and M. Turner, eds. 2008. *Whose Peace?: Critical Perspectives on the Political Economy of Peacebuilding*. Basingstoke: Palgrave Macmillan.

Richmond, O.P. and R. Mac Ginty. 2015. Where Now for the Critique of the Liberal Peace? *Cooperation and Conflict*, 50(2), 171–89.

Robinson, J.A. 2014. ¿Cómo modernizar a Colombia? *El Espectador*, 14 Dec.

Robinson, J.A. 2015. Colombia: ¿esta vez es diferente? *El Espectador*, 18 Jan.

Rosanvallon, P. 1999. *Le capitalisme utopique. Critique de l'idéologie économique*. Paris: Seuil.

INDEX

Note: numbers in **bold** refer to Tables

For Product Safety Concerns and Information please contact our EU representative GPSR@taylorandfrancis.com
Taylor & Francis Verlag GmbH, Kaufingerstraße 24, 80331 München, Germany

www.ingramcontent.com/pod-product-compliance
Ingram Content Group UK Ltd.
Pitfield, Milton Keynes, MK11 3LW, UK
UKHW021307260926
475702UK00048B/810
9780367469634